The Shape of Future Technology
The Anthropocentric Alternative

The Springer Series on

ARTIFICIAL INTELLIGENCE AND SOCIETY

Series Editor: KARAMJIT S. GILL

Knowledge, Skill and Artificial Intelligence
Bo Göranzon and Ingela Josefson (Eds.)

Artificial Intelligence, Culture and Language:
On Education and Work
Bo Göranzon and Magnus Florin (Eds.)

Designing Human-centred Technology: A Cross-disciplinary
Project in Computer-aided Manufacturing
H. H. Rosenbrock (Ed.)

Peter Brödner

The Shape of Future Technology

The Anthropocentric Alternative

With 23 Figures

Springer-Verlag
London Berlin Heidelberg New York
Paris Tokyo Hong Kong

Peter Brödner
Wissenschaftszentrum Nordrhein-Westfalen
Institut Arbeit und Technik
Florastrasse 9
D-4650 Gelsenkirchen 1
West Germany

British Library Cataloguing in Publication Data
Brödner, Peter
The shape of future technology: the anthropocentric alternative. – (Artificial
  intelligence and society)
1. Technology. Social aspects
I. Title
306.46

Library of Congress Cataloging-in-Publication Data
Brödner, Peter
[Fabrik 2000. English]
The shape of future technology: the anthropocentric alternative/
Peter Brödner. – English ed.
    p.   cm.
Translation of: Fabrik 2000. 2. Aufl., 1986, c1985.
Includes bibliographical references.
ISBN-13: 978-3-540-19576-4   e-ISBN-13: 978-1-4471-1733-9
DOI: 10.1007/978-1-4471-1733-9

1. Production management.   2. Automation.   I. Title
TS155.B67613   1990   90-9653
670.42–dc20   CIP

First published as *Fabrik 2000: Alternative Entwicklungspfade in die Zukunft
der Fabrik* by edition sigma rainer bohn verlag, Berlin 1985
Reprinted 1986
Second revised edition 1986
English edition 1990

2128/3916-543210   Printed on acid-free paper

Acknowledgements

The present work was written between January and March 1985, during my stay as a guest at the Science Centre of Berlin, International Institute for Comparative Social Research/the Politics of Work. I would particularly like to thank:

the Science Centre of Berlin, and in particular Prof. Dr. Frieder Naschold for his kind invitation to stay,

those working at the Institute for the Politics of Work, for their many stimulating conversations,

the Atomic Research Centre at Karlsruhe, in particular Dr. Dietrich Stams and Dr. Tomas Martin for their kind support for the project,

my colleagues at Project Supporter Production Engineering for bearing the burden of work, in solidarity, during my absence,

Frau Alter for quickly drawing up the manuscript, but without detriment to quality.

Karlsruhe, August 1985   Peter Brödner

Contents

Introduction

Mike Cooley

One of the most remarkable features of modern industrial society, is the gap between that which technology *could* provide for society (its potential) and that which it actually does provide for society (its reality). We have for example, complex control systems which can guide a missile to another continent with extraordinary accuracy, yet the blind and the disabled have to stagger around our cities in very much the same way as they did in mediaeval times. There are advanced communication systems enabling messages to be sent around the world in a fraction of a second, but it now takes longer to send an ordinary letter from Washington to New York than it did in the days of the stage coach.

Such a growing chasm between potential and reality, is giving rise to a thorough questioning of many of the orthodoxies in these areas and the priorities on which they are based. Similar contradictions, even if at this stage less obvious and dramatic, abound in the field of manufacturing technology. There, we find technologies which have the potential of liberating human beings from soul-destroying, routine, backbreaking tasks and leave them free to engage in more creative work, but which in reality, often end up reducing the human being to a mere machine appendage, acted upon by the technology and becoming a passive, pathetic element in the productive system rather than a creative, dynamic human being.

Many of these technologies result in dramatic deskilling. It has for example, been reported in the "American Machinist" [1] that for some applications of NC technology, the ideal workers are mentally retarded, and a mental age of twelve is advocated. Had the objective been to provide work for the mentally retarded, this effort would have been laudable. However, what we were in fact witnessing was the destruction of the very seedbeds from which the future generations of skill and knowledge grow. Such consequences arise partly from the manner in which we view the "scientific design" of systems in general, and in this case, production systems. We seek to ensure that they display the three predominant characteristics of Western science and technology; namely, predictability, repeatability and mathematical quantifiability. Those by definition preclude intuition, subjective judgement, tacit knowledge and intentionality, which are among the key elements displayed by skilled workers.

Furthermore, we tend to regard a good systems design as that which reduces or eliminates uncertainty. Since however, systems designers perceive a human being as constituting an uncertainty, it follows with some kind of Jesuitical logic that a good design is one which marginalizes human intelligence. More perceptive employers are now beginning to realize that in the long term this may be counter productive. They are beginning to understand that one should not be thinking merely in terms of production, but the reproduction of knowledge. From where they ask, will the next generations of skill and knowledge come if people are denied the space within their productive work, to develop new skill and competence?

In addition to this, there are beginning to be even deeper concerns. If systems are highly synchronized and machine dependent, then when one part of the system goes down, the high-level synchronization is suddenly turned into its dialectical opposite, and there is high-level desynchronization in an almost catastrophe theory sense. If however, within production, there are highly skilled, proactive, competent workers for whom the process is transparent, they are capable of dealing with wide bands of uncertainty, and the result is a system which may be said to be more robust in general systems terms.

The economic dimension of this, is already beginning to be self evident. In a recent report comparing downtime between systems in the United Kingdom and the Federal Republic of Germany, it was found that the downtime was significantly greater in the UK. Initially, it was felt that this might be due to more antiquated equipment. However, when like with like was compared, it was still found to be the case that the downtime was significantly greater in the UK. Deeper analysis revealed that this was because the Germans were using highly skilled, proactive "Facharbeiter", who were capable of anticipating systems failure and taking remedial steps before it became critical or, if part of the system did fail, they were frequently able to restore it to its operating condition. Failing that, they knew precisely whom they should call upon to deal with a particular part of the systems failure.

In spite of this however, the tendency worldwide still seems to be to attempt to replace human beings by increasing levels of fixed capital. It has been pointed out for example, that in its attempts to compete with Toyota, General Motors engaged in a massive programme of re-equipping based on very advanced manufacturing technology. Indeed, it is said that the level of investment was so great that they could have bought Toyota instead [2], yet they are still incapable of competing with it because the Japanese sought to draw much more on human skill and ability. Indeed, in some of their quality circles (and I say this without in any way advocating them), they do talk about getting at "the gold in workers' minds".

However, even in the United States, this precarious dependence on machines is beginning to be questioned, and the value of skilled and capable people is beginning to be reassessed. Thus, in a recent report by a distinguished group of experts on US industrial performance, the following is pointedly stated.

"The American industry of the 50s and 60s, pursued flexibility by hiring and firing workers who had limited skills rather than relying on multi-skilled workers. Worker responsibility and input progressively narrowed, and management tended to treat workers as a cost to be controlled, not as an asset to be developed". [3]

Peter Brödner's book highlights these issues. His method of analysis and the cases he cites are located within the German industrial tradition. Few observers would now deny that the extraordinary success of the German manufacturing industry and its export performance, arises in large part from the vast reservoir of skill and knowledge that employers are able to draw upon. This results from a well structured, apprenticeship-type scheme which produces a highly skilled and talented workforce. Furthermore, promotion to higher levels of production engineering activities is available for those who wish to progress.

Peter Brödner, who has a deep technical knowledge of production technology, has played a significant part in the indicative planning of sectors of German industry through his work in Karlsruhe. He is therefore, extremely well placed to draw these matters to the attention of the international engineering community, and a great advantage of the present book is that it makes available, to those not intimately acquainted with the structure of German industry and its range of skills, some of the pivotal arguments now being pursued in that country. This questioning extends more widely of course than Germany, and is beginning to be evident throughout Europe as a whole. In the early 80s, a group of researchers at Manchester University with Professor Rosenbrock, worked on a Human-centred Manufacturing System and a fascinating account of the multidisciplinary design team's work is available [4].

Increasingly, it is being recognized that one should not accept that there is merely one type of manufacturing technology, namely, the American Fordist, Taylorist form. Ten partners in three European countries came together in 1984 and pointed out to the EEC that technology should be perceived as part of culture, and just as cultures produce different languages, different music and different literature, so there should also be different forms of technology emerging which reflect the cultural, educational, political and ideological realities of the countries in which the technology was to be implemented. They argued for a European form of technology, in which those European aspirations (frequently expressed more in the rhetoric than demonstrated in practice) of freedom of the individual, self-consciousness, capacity to develop and a sense of quality, should be reflected in the design of the technology.

The result is ESPRIT Project 1217 whose aim is to design and build the world's first Human-centred, Computer Integrated Manufacturing System [5]. The outcome of that project so far, begins to reflect in part some of those changes advocated by Peter Brödner in this present book. In the system, the worker deals with the qualitative subjective judgements and the machine merely deals with the quantitative elements. At the production planning and scheduling level, the object is to render processes highly transparent. At the computer-aided manufacturing (CAM) level, the deep tacit knowledge of the skilled worker is accepted as valid scientific knowledge, and a variety of shop floor programming techniques are made available, including graphical input. At the computer-aided design (CAD) level, the usual problem is that of menu-driven systems, in which the designer can make pleasing patterns of predetermined elements but has difficulty in changing those elements. To confront this, an electronic sketchpad has been developed, which allows the designer to create any elements he or she wishes in a rough form. Furthermore, the portable electronic sketchpad facilitates a dialogue, either directly or electronically, with people on the shop floor, so that the vast experience of those in the workshops is reflected in the design before that design is consolidated.

Systems of this kind all support the wider concept projected by Peter Brödner in this book, that of anthropocentric systems. Indeed, a panel of experts has prepared a report for the FAST programme of the EEC entitled "European Competitiveness in the 21st Century: The Integration of Work, Culture and Technology" in which they advocate the introduction of anthropocentric systems and propose some 200 million ECUs be provided for research programmes to develop the various subsystems over the next three years [6].

It is beginning to be recognized that the existing systems display very high noise, but frequently a low signal. Data systems are characterized by high noise. If the data is so organized as to transform it into information, the noise reduces and the signal becomes stronger. If the information is gradually applied it becomes knowledge and if that knowledge is absorbed within a culture it begins to be wisdom and wisdom should lead to informed action. It is at this knowledge/wisdom/action end of the cybernetic spectrum, where there is little noise and where the signal is strong, that these systems should be designed [7]. This makes full use of the most precious asset a company has, which is the skill, ingenuity and creativity of its people.

Peter Brödner rightly points out that some employers, in order to avoid dependence on human intelligence, are increasingly hoping that expert systems will solve their problems. It should be said in passing, that most systems tend to be expert replacement systems rather than expert systems as such. Peter Brödner's critique of these systems, and their related areas such as artificial intelligence (AI), makes compelling reading and challenges many of the assumptions of the AI community, one of whom recently said "Human beings will have to accept their true place in the evolutionary hierarchy namely, animals, human beings and intelligent machines". This could indeed become a self-fulfilling prophecy in the short term at least, unless we begin to provide tools for people in the sense in which Heidegger distinguished between a tool and a machine [8].

Peter Brödner, raises very directly, the issue of accountancy systems. For too long in the United States and the United Kingdom, manufacturing industry has been dominated by the shortsighted perceptions of its accountants. This in part, is a reflection of the overall dominance of the short-term value system of finance capital as distinct from the more long-term developmental attributes of industrial capital which is evident in West Germany and Japan. Even the accountancy community itself, is beginning to question its own criteria in these matters. Thus Professor Tony Hopwood, a Professor of International Accountancy at the London School of Economics, pointed out that cost accountants in an industrial setting, spend about 60% of their time seeking to reduce labour costs, when those labour costs constitute only about 12% to 15% of the total costs in any case [9].

The book also raises in a very direct way, the reality that whether we go for anthropocentric systems or for the conventional ones, there will be a reduction in the numbers employed in manufacturing, given the various constraints. Clearly, this is a key issue, and although it is outside the scope of this work, he does point out that the metalworkers' union is attempting to address these issues and increasingly, both in Germany and elsewhere, the concept of the shorter working week as a form of creative job sharing, has begun to be linked with the idea of opening up social markets in which products will be produced that are socially useful and environmentally desirable. Thus, one could begin to see that the productivity increases which these technologies afford, could gradually be transformed into a wider social good.

Alongside this, are those organizational forms which reinforce control and power relationships. There are political, socio-technical and ideological considerations to take into account. Peter Brödner highlights these contradictions vividly. He reveals on the one hand the inefficiencies of Taylorist forms of production and on the other, provides statistical and economic evidence to support his searing critique of the existing forms of manufacturing technology. He contrasts mass production with workshop production, and he challenges head-on, the almost universal assertion "that modern production systems create multifaceted and responsible jobs requiring correspondingly intense training". He reveals vividly, the drive toward the workerless factory, no doubt to be followed by the workerless office. He also analyses the impact of automation on intellectual work and knowledge-based systems.

This book however, is not the usual "critical study", central to so many books on the impact of technology. This book is far more, for Peter Brödner succeeds in transforming a sociological critique and economic analysis into a series of technological proposals and that is quite unique among writers on these topics. Equipped as he is, with a deep technical understanding of production technologies through his work at the Production Engineering Centre at Karlsruhe, he is capable of bringing to the topic a professional understanding which makes his analysis and proposals very convincing.

He highlights the contradictory nature of "technological progress" and his analysis suggests that changes in the form of work may be necessary, not only from a humanitarian point of view, but also for very sound technological and economic reasons. His approach is somewhat dialectical. Thus he sees the contradictory nature of these developments and the inadequate strategies for dealing with them as a unique historical opportunity to shift from a technocentric form of production to an anthropocentric form. The issues he raises will be of vital importance, not only to engineers and systems designers, and of course industrial sociologists, but should also in my view, be of primary concern for trade unionists.

For too long, trade unions have regarded the development of production technologies as being essentially neutral, and they have perceived the problem as capitalism's misuse of these technologies. The present book reveals a much richer analysis of the problem, and could provide an industrial relations framework in which trade unionists move from reacting to the proposals of the multinationals for technological change, to a much more constructive and proactive one in which they assert the right to have systems which build upon the skill and ingenuity of their members.

The book will also help to establish a correct balance in the political approach to these topics. For too long there has been an obsession with the contradictions of distribution at the expense of a serious ongoing analysis of the contradictions of production. Peter Brödner's book is a controversial and challenging one, and not all its assumptions will be accepted by those who generally support the thrust of Brödner's reasoning. There will be some concern that there appears to be an accepted division between the design area of production and the manufacturing area. It may be possible, using advanced computer-based systems, to bridge that gap, and move forward towards a more "design by doing" situation. The thrust of the book however, is precisely that it brings these issues into clear focus, and provides us with an aperture through which to view a whole series of contradictions within manufacturing technology.

The appearance of an English edition is to be particularly welcomed. It is essential that topics of this kind are discussed worldwide and now, readers in Australia, America, Britain and throughout the English-speaking world, will have the opportunity to acquaint themselves with a very important discussion now taking place in Germany. The extensive reference list will also help readers to become acquainted with the literature in this area, although it is inevitable in a book of this kind that some of the material is in German.

A translation, however well done, never fully succeeds in doing complete justice to the original. This is particularly so where a book deals with issues within a tradition, that is, within the context of specific educational, political, social and even industrial developments. At the most simple level how, for example, should one translate "Meister" when there is no such grade in UK or US industry? The word "foreman", used here, is not entirely satisfactory.

In general, we have sought to convey the original framework of argumentation, although, for example, "capitalist production" sounds somewhat more shrill and polemical than in its German form. Furthermore, the line of reasoning and terms used in those sections dealing with political economy and "work science" may be unfamiliar and sound somewhat convoluted. Concepts such as "objectivized knowledge" and "alienated work" may appear remote to some English-speaking engineers. However, this is but a small translation-price to pay for access to a book which so richly encapsulates a point of view which has already had a great impact on the outlook of many leading industrial engineers and companies in Europe's most successful manufacturing nation.

It is to be hoped that the analysis advanced in this book will provide one of the intellectual and political building blocks necessary in the process of constructing a form of technology which enhances human skill and our environment rather than diminishing it. Given the tremendous rate of technological change and the infrastructure now being put into place which may permanently close off options for alternatives, the book should provide a powerful impetus for positive action. Peter Brödner would agree I am sure, that the issue is not just to analyse the industrial world, but rather to change it!

References

1. Cited in Cooley, M.J.E. 1987 *Architect or Bee? The Human Price of Technology*, Chatto & Windus/Hogarth Press, London, p. 130
2. Cannon, P. "The human factor returns", *The Independent*, 10th October 1988, p. 18
3. Berger, S. et al. 1989 "Towards a new industrial America", *Scientific American* **260**(6) p. 40
4. Rosenbrock, H.H. 1989 *Designing Human-centred Technology*, Artificial Intelligence and Society Series, Springer-Verlag, London
5. Cooley, M.J.E. op. cit. pp. 147–154
6. FAST Report 1989 "European competitiveness in the 21st century: The integration of work, culture and technology", Commission of the European Communities, DG.XII/H/3/FAST, Brussels
7. Cooley, M.J.E. op. cit. p. 12
8. Ehn, P. 1988 "Work oriented design of computer artifacts", Arbetslivcentrum
9. Hopwood, T. 1989 "Production and finance: The need for a common language", *Proc. New Manufacturing Imperatives*, Axiom Systems Design Ltd, 1 Logan Mews, London W8

Chapter 1

The Factory of the Future

1.1 Prophecies by the Dozen

Rarely, in the history of capitalist industrialization, have such diverse scenarios for the future development of the factory been put forward. In no other industrial age have the ideas been so far-reaching and bold, or the slogans so provocative and challenging as in the current debate on the future of the factory. Never before has a discussion about technology and the deployment of labour in the productive process assumed the character of a crusade.

However, it isn't just the design of the factory of the future that is in question. Present forms of production technology and work organization are on the agenda as well as their future development. These involve a high level of investment and high risk taking. The stakes are high in the current race to restructure, since anyone who makes a wrong investment as the result of a misguided view of production itself, can easily be ruined. This is just as much the case in respect of state support for research and development as it is for individual businesses.

What then is this argument all about? What plans and what standpoints do the competing parties adopt? On one side of this argument we have advocates of a technocratically narrow minded production concept (Kern and Schumann 1984). They have written on their banners the captions "automated factory", "unmanned factory" and "computer integrated factory". These are all attempts to define the "factory of the future".

This final slogan is moreover, an example of what George Orwell accurately termed "newspeak" and it may be presumed that there are compelling reasons for using it. After all, the term "workerless factory" which was so triumphantly proclaimed at the initial stages, could arouse the suspicion and resistance of trade unions in times of continuing mass unemployment. It is thus more socially acceptable to talk about the "factory of the future", for who indeed, could fail to have faith in the future?

The expression however, signifies even more than that. When these people refer to the "factory of the future", what they are really doing is supporting their own position. Even so, one thing is not in dispute. The factory as such does have a future, but what kind of future? In the present situation it will be tied to the dogmas of technological progress and technological determinism.

The beliefs embodied in this outlook were summarized some five years ago in a publication (Spur 1984) which has proven to be the most important recent exposition.

Almost without the public noticing it, we have crossed the threshold to the new generation of factories. Productivity, speed of production, flexibility, quality and reliability will reach a level unattainable through conventional production techniques. The highly differentiated functions and technological refinement of industrial products lead to increased product variation, and the rapid substitution of one product for another. At the same time, rising labour costs, together with the requirement for flexibility and high productivity, are factors which militate in favour of automation. The consequence of all this, is the creation of the computer integrated factory.

Computer integration is at the core of future production innovation. It will lead to automatic production by means of variable product programmes, and control the continual optimization of the flow and processing of materials as well as the dynamic configuration of all the means of production. Rather like an organism consisting of machinery with intelligence programmed in and stored, it will be able to generate new production configurations for producing goods automatically. In this higher stage of development, the factory will need machine intelligence. (Spur 1984, pp. IVf)

The complete dominance of this standpoint in certain specialist publications, conferences and universities must not be allowed to hide the fact that within the factory itself, the attitude has not remained unchallenged. After all it is the factory itself whose future is at stake and nowadays it is ever more common to meet supporters of an "empirical, non-ideological view of production" (Kern and Schumann 1984) who, with key phrases such as "the anachronistic factory" (Williamson 1972), "more productivity through less division of labour" (Moll 1983) and "production islands: an alternative work structure" (Ahlmann 1980), are looking for new responses to the present challenges. This counter-position which they have taken up has itself been formulated no less prominently, and no less fundamentally elsewhere.

Technological change in manufacture is inseparable from the people who are a necessary, even if decreasing, part of the manufacturing process, and many of our present difficulties have been caused by considering changes to the mechanics of the process in isolation from the people who have to make them work. The blame for this falls also on the scientist and the engineer, who, by their mechanistic concepts, initially in the relationship between man, machines and processes, and continued more recently in the field of computers and information handling, have encouraged a belief in the superiority of the machine. We seem to be encouraging a ludicrous vision of industry where computers will do the decision-making, while men provide low-grade motive- or brain-power to carry out increasingly simple and repetitive tasks. A little reflection reveals the ultimate lunacy of such a proposition, and it can be recognized for what it is, a misuse of technology, and a misunderstanding of the proper functions of man and machine. (Williamson 1972, pp. 139f)

The alternative production structure can be seen in group technology. Its basic principle is that:

it consists of pockets of self-contained responsibility in which man's skill, intelligence and enthusiasm are harnessed in somewhat specialized working groups, which can extract the best possible result from the level of manufacturing technology in the context of the particular circumstances which apply at the time. (ibid. p. 152)

The contradictions between competing statements and visions of the future have not arisen by chance; nor has their proliferation. A situation of radical change is being expressed here. This situation is further characterized by the variety of social standpoints and by economic factors including the decline in industrial growth alongside the sharp increase in capital intensity, and the lasting and far reaching changes in the world market situation.

There is now a new international division of labour which has come about through the export of industrialization to the emerging countries, and from the sharper competition in the industrialized countries themselves, including of course, the progress of advanced technologies. All this has contributed substantially to the unfavourable climate in the world market.

It is predicted that low levels of production will continue for a long time yet, and it is unlikely that demand can be met with conventional production structures. The large number of competing suggestions for redesigning production processes, and the vehemenece with which these views are being put forward, indicate that the time is ripe for a change in production theory, both in the technological and organizational fields. The factory of the future is at a crossroads.

1.2 Banish the False Prophets!

But what is the basis of these vastly differing outlooks, and how can they be evaluated? How can true prophets be distinguished from false prophets? How should production be designed – both technologically and organizationally – in order to endure the long stormy journey ahead, and not to capsize on the way? The main purpose of this investigation is to gain a clearer understanding of the conditions and possible consequences of certain plans for the development of the factory, and to tentatively put forward satisfactory answers to the questions raised.

If this is not to be done simply through intuition, where the outcome has to be either accepted or rejected, than there has to be a solid basis of theoretical understanding and an analysis of empirical findings concerning the origins of the factory of today and the conditions under which it is developing, in order to conduct the discussion. If however, the intention is to look into the future with some degree of certainty, despite discontinuities, then we have to be clear about the past. This cannot be done properly without discussing traditional dogmas. These are the ideas centred around the present and future relationship between manufacturing, the deployment of labour and the interests of society.

The dogmas of technological progress and technological determinism have held the field in the discussion for a long time. The high priests of academia proclaim them, politicians act in accordance with them (in matters great and small), and the public pays homage to them. According to these dogmas, technological development proceeds initially of its own accord, spurred on by its own logic which in principle cannot be influenced. The technology used in production then determines the way the work is organized. In consequence, it also determines the demands made on the labour force for certain skills and abilities. Any flexibility in the design of this work organization would thus be impossible. "For better or for worse", thus runs the slogan of the Club of Rome (Friedrichs and Schaff 1984).

This slogan is akin to the position taken up by the Federal Republic's Council of Experts for the Evaluation of Economic Development (known as the five wise men). We are therefore condemned, with our export dependent economy, to compete in the race to be the top performer in technology and the top performer in its application to production. If not, we collapse as an economic force.

Industrial sociology too, cottoned on to technological determinism a long time ago, albeit with some differences in detail. It formed part of the interpretive framework of its investigations (see e.g. Popitz et al. 1957; Bright 1959; Blauner 1964; Mallet 1972; Gorz 1980). However, these investigations have not remained unchallenged. Since then, sufficient theoretically based and empirically sound data has been collected to undermine the key tenets of technological determinism and hence place it in the realms of mythology.

Three specific research approaches which have contributed substantially to this are outlined below:

Firstly, there is the plant-oriented approach (as specifically described by Altmann and Bechtle 1971). This attempts to clarify how technological and organizational changes in production come about under the influence of activities in a plant which is striving to maintain its independence (see e.g. Lutz and Schultz-Wild 1983).

Secondly, there is the study entitled "New Technologies and Alternative Forms of Work" (Benz-Overhage et al. 1982) which attempts to interpret, from the viewpoint of technology and the organization of labour, the restructuring strategies and developmental trends in areas of industry where production and the requirements of the market are in conflict. It also attempts to clarify the particular functions of computer technology in the organization of production processes in the areas discussed, and indicates their application in the design process.

Thirdly, there are several variants of the approach based on the politics of work (see e.g. Edwards 1981; Sabel 1982; Naschold 1984) which take the basic issue of the politics of work in action (transforming the ability to work into actually doing it) as a starting-point for clarifying how the effort to secure control in the factory, influences the way the technology and organizational structure of the production process is organized.

The study outlined here adopts the essential ideas of these research approaches, and is set in a theoretical foundation consisting of three basic transformational relationships in the capitalist production process.

1. The conversion of surplus value into profit (circulation)
2. The conversion of labour power into actual work (surplus value production)
3. The reproduction of work by objectivized knowledge (material production)

Concerning (1)
The aim of the capitalist production process is the production of surplus value. However, to the great annoyance of those involved, this can only take place under conditions which cannot be monitored by the individual company. Firstly, its production is tied to the manufacture of goods using human labour, and secondly, these goods have to be put on the market in appropriate quantities and quality, at the right time and at the right price, in order to find buyers. This is the way the demands of production and of the market economy are formed under the pressure of competition. The market economy governs the relationship between quantities, quality and prices of the saleable products, and this determines how capital is used in the circulation process.

The production economy supplies the design principles for production, and the production process is subject to the conditions under which capital is deployed, in accordance with these principles. Regardless of relationships (2) and (3), the production economy and market economy substantially determine the organization of labour, the techniques of production and the synchronization of the separate processes. It follows then that they also establish costs, times taken and the way in which human labour is subordinated under capital.

In the production process where there is division of labour, and co-ordination by means of an overall command structure, socialized labour becomes productive within the context of the market economy. However, since the various production processes are independent of one another, the demands of the production economy contradict those of the market economy. In other words, a pre-defined production set-up can be too inflexible to be capable of reacting appropriately to market demands, or cost structures may necessitate a sales strategy which has become incompatible with market demands (Benz-Overhage et al. 1982).

Concerning (2)
Once a contract of employment has been agreed, management has at its disposal the labour power which it has purchased as a means for the extraction of surplus value, but this does not as yet determine the precise physical form of the work by which this highly versatile unit of labour power can, by himself, become an effective means of production. However, labour power cannot be separated from its owner – the person at work – since physical and mental attributes are combined in the one individual. As a consequence, whoever has purchased this labour power cannot have sole control over it; so to make use of it he must ensure the co-operation of its subject. He thus is also provided with the self-will of human labour and therein is the core of his problem.

To thoroughly transform labour power into productive work, management has to govern and control within the factory. Among the many vehicles for doing this, the techniques of production and the organization of labour are the most effective. Consequently, the configuration of the production process is determined by the necessity to ensure control over the workforce in the factory (Noble 1979; Edwards 1981; Dörr et al. 1984).

Concerning (3)
It is only in respect of the human being in the work process (independent of the type of production) that the tasks performed when working have already existed in his mind as a plan of action. The varying patterns of action are created on the basis of the human being's internal representation of the activity itself, and also of the change in the environment that will be caused by it, both of which are absorbed from past experience. In some circumstances, which still need to be defined precisely, the worker is able to objectivize the knowledge gained through personal experience, into patterns of activity related to language, tools or even machines. These patterns are handed on to other individuals or even to future generations, on whose actions they make demands (Volpert 1984a).

With the assistance of such ideas, theories and models of the production process can be constructed, but since the patterns referred to have been obtained by abstracting from past experience in each individual case, it is inevitable that they will reflect that experience in an incomplete and patchy manner. Now, an essential element of management control, is to extract this knowledge of production from those who work, and present it back to them in an objectivized form, as an external force; that is, in the form of orders, procedures and machines. Seen in this way, machines are nothing more than the implementation of production process theory (Bednarz et al. 1984). However, the incompleteness of such objectivization and the procedures derived from it, lead to discrepancies between the actual production process and its model that cannot be reconciled, and which place objective limits on control.

As will have become evident already, these fundamental relationships do not exist in isolation. Rather, a contradiction exists between them. For example, the requirements of capital (1) may place demands on the flexibility of the production process which are irreconcilable with the existing control structures (2) – structures originally set up to gain the upper hand in the factory. The imperatives involved in ensuring this control (2), lead to working conditions and regulations which actually hinder the workers in their attempts to contribute the knowledge they have gained from experience. Not only that, they prevent them from correcting actual mistakes (3). Indeed the discrepancies between the work process and its model may be so great, and the objectivized knowledge so inadequate, that market opportunities for deploying capital cannot be exploited.

It will become clear later in the book that these relationships provide an adequate framework in which to understand the determining factors in the development of capitalist production. They are suitable for the theoretical reconstruction of past development, and they are particularly suitable for understanding the current state of radical change, for forecasting the opportunities in design and for assessing options for future development.

These relationships do not of course, appear obvious in individual firms. In fact the nature of the forces which determine events are effectively piecemeal, due to certain internal and external constraints. In each case the business strives for independence. Affecting this are the practical decisions on organization, use of personnel and use of investments. These of course depend on the range of products, the nature of the production and the skills and abilities demanded of the workforce. They also depend on social relationships, company policy and on supply and demand in the markets (Altmann and Bechtle 1971). Only this can explain the extraordinary profusion of production processes existing.

The connection between technology and the organization of work in particular, reveal clear similarities and identical development within for example, manufacturing, where standard principles are applied. The intention of the present study is to discover common features within a framework of variety.

At this point, it is useful to highlight the fact that industrial production always involves subjective, non-verbal communication, and thus it represents a social process undergoing lasting changes as a result of the dynamics of the transformational relationships (1), (2) and (3) described above. In the course of these changes special forms of communication develop on a material (non-verbal) level, via the technology and as a result of the work organization. Beyond this, the changes produce certain kinds of sanctions on the part of management, and forms of solidarity such as helpfulness and comradeship on the part of the workforce. These result in the particular attitudes and expressions, based on a shared understanding of the work situation, which give forms of communication their special character. Unfortunately, we will have to forego here, the study of this microcosm of social relationships within the factory and its associated agencies and not give it the attention it deserves.

First of all then, we will outline the essential features of capitalist restructuring against the background of its social roots. What caused employers to divide work performed entirely by skilled manual workers first horizontally (Smith 1776; Babbage 1835), replacing it in part by machines, and then vertically by separating planning from implementation (Taylor 1911)? It now seems feasible to automate intellectual work as well. How can this even be a possibility? How and where is this happening and what are its limits? The answers to these questions must be found, so that they can be used to trace two alternative developmental routes leading to the factory of the future, the technocentric and the anthropocentric. They will be compared as two opposing ideals for overcoming current difficulties, and they will be judged according to their potential.

Before this is done an initial preliminary conclusion can be arrived at here. It is evident that in manufacturing, the basic relationships and their interconnection are in fundamental contrast to the myth of autonomous technological progress and technological determinism, whose central deficiencies can be seen even at a first glance. The ideas underlying them cannot provide an explanation for the observed fact that this same technology can be incorporated in widely differing forms of labour organization, and the outcome can be equally efficient economically in every case.

Nor can these ideas provide an explanation for the way the autonomous laws of technological development are formed; this development being itself, just as obviously, the result of human work. Are they supposed to spring from intuition? or from the "Faustian urge" of inventors' minds? It is not technology that determines the production process and the way work is organized. It is social circumstances which give rise to the conditions and the motivation for determining the development and use of technology and the organization of labour. So long as these ideas are given credence, they are extremely convenient politically, to put into use. For since technological development cannot be influenced anyway (so the argument goes), any discussion about it is unnecessary. It is sufficient to promote its acceptance and absorb its social consequences.

Even within the technological sphere as such, it is frequently possible to simply limit oneself to the adoption of the very latest developments. It will only be tiresome and costly if one is the first to do it. This approach, that of imitating others and forgetting about the special circumstances in which they achieved success, rather than building on one's own strengths is extremely dangerous. This is especially true in situations of radical change. A example of this is the disaster which resulted from the policies applied in the development of German main-frame computers, despite the fact that the manufacturers were warned against their attempt to undermine the market leader's position by imitating its products.

The belief that technological progress will unswervingly follow its own route and that its consequences are irrevocable is a myth suited to a capitalist situation. It is suited in two ways. Firstly, it has a track record as being useful and secondly, spurred on by competition, it internalizes the movement of capital in the form of objectivized knowledge. It is therefore, the ideological expression of the accumulation of capital. What appears to be a technological necessity is in fact only an ideological necessity. We will frequently encounter such beliefs further.

1.3   The Workshop, not the Production Line

Product manufacturing processes are the only ones under consideration in this book. The conversion and distribution of energy, and processes involving materials alone (for example in the chemical and steel industry), are not addressed here.

In the production of goods with a definitive geometrical shape, two separate production processes have developed which exhibit widely differing forms of production engineering and labour organization. In the production of individual items and in small-series production usually linked to actual orders, work is done on the basis of routine tasks where standard processes are combined in a spatial and organizational sense (workshop production), and in the main, general-purpose machines are utilized. Mass production, in which the client remains anonymous, is organized on the production line principle, with interlinked specialist machines.

In the past, where products were required in very large quantities, market demand encouraged the production line manufacture of large numbers of identical items, inflexibly automated to a large extent (in the form of transfer production lines for example).

Workshop production, which is very costly, has always baulked at automation since the extremely large variety of parts and products required demands a high degree of flexibility. Just as new market demands for greater variety in mass-produced goods have sparked off huge efforts in developing new designs for mass production, so the costly and time-consuming flexibility of small-series production has invoked many attempts to automate it.

The efforts in both these fields have made use of computer technology which can be used in all situations, rendering more flexible the hitherto rigidly automated mass production process and in addition, automating the production of individual items. As a result, paths of development that were at first divergent are now exhibiting common features. Nevertheless, fundamental differences remain. The products of mass production, whether produced in large batches or linked to orders, are always fully defined, and the process for their production is, without exception, planned before the first item is produced. If the products are made to order, then they have to be individually designed when the production takes place. There is a limit therefore to the extent of advance planning in this type of production.

The concern in this book is mainly with the production of capital goods. These are almost exclusively produced by one-off and small-series production linked to customer orders and as they make up nearly a quarter of the industrial production of the Federal Republic of Germany, they form the backbone of its national economy. Along with the chemical industry, we owe them thanks for our rise as an industrial nation. Moreover, high exports expose us to increasing competitive pressures worldwide.

Mechanical engineering alone, which is a central sector of the capital-goods industry, produced goods to the value of DM 113 000 million in 1981, of which 61.5% were exported. The proceeds from its exports alone paid off 92% of the oil debt. Intensified by the crisis of recent years, but not brought about by it alone, the standing of this industry in the world market has become somewhat uncertain. Production problems signifying radical change are constantly being highlighted. It is nevertheless worth looking over the fence occasionally, at the car industry, a traditional branch of mass production. The search for new modes of production can be seen there as well (Kern and Schumann 1984), allowing us to make instructive comparisons concerning human labour.

Chapter 2

The Origins and Nature of the Factory

2.1   The Heroic Phase

2.1.1   Increased Production through Division of Labour

Within distribution, trading capital has created for itself an institutional form which enables it to appropriate skilled manual production and the trade in its products and to acquire the resulting surplus value.

In the past, the distributor delivered the raw materials to the producers, who were craftsmen scattered about in the surrounding rural areas, and collected the finished or partly finished products, having previously agreed the quality, quantity and price in order to market them. As a third party, he separated the producers from their customers, and thus excluded them from the circulation process (the conversion of surplus value into profit). By this means, he had them so well under control that he could reimburse them on average, just enough to put them in a position to reproduce their labour power, and then he was able to induce them to produce so much that the proceeds he obtained from the sale of their products far exceeded the expenses he had incurred for wages and materials. In this way, the distributor was able to appropriate from the producer, the difference in value between the working hours spent on producing the product and the working hours necessary to reproduce his capacity for work (labour power). In other words he appropriated the surplus value as a profit.

But, advantageous as this was for the distributor at the outset, there was of course a snag. As a transitional form the system was extremely unstable. Although the distributors had relied on production, they left the production process unchanged and were completely dependent on the ability and judgement of the skilled producers. Spurred on by their desire for profit, they had no option but to make the producers manufacture more and more products. The interests of the producers, however, were quite the reverse.

Higher wages were no real solution from the distributor's viewpoint because then the producers could produce less and enjoy some leisure time while still making a living. Keeping down wages didn't help either because this made the producers resist the system by withholding raw materials and supplying inferior quality products. Moreover, the distributors could never put a stop to attempts by the producers to "freelance" in secret using embezzled materials, or to break out of the system completely. The Draconian legal punishments threatened for such deception show the contradictions inherent in this system and the tough struggles it engendered (Marglin 1977; Kuby 1980). The only solution now, was to bring the production process under the umbrella of manufacturing, and to co-ordinate and monitor it. After all, higher and higher transport costs were being incurred in distribution as a result of expanding markets. To overcome once and for all, the contradictions in the distribution system and the causes of its instability in manufacturing, drastic structural changes had to be made.

Since the producers were not ready to give up their customary living conditions voluntarily, they were compelled to do so in the course of radical social change which was both lengthy and painful and during which they were dispossessed of the means of production by capitalism. One way was to use agencies which had power, to enforce the privatization of common land used by smallholders; another was to impound working materials and partially made products from the craftsmen who had run into difficulties due to the lack of orders. The producers had been robbed of their livelihood and so, in the end, they were forced to put their labour power, their only remaining possession, on the market. By doing this, they could earn a wage which would keep body and soul together.

When the new industries started up, work was carried out with the same materials and in exactly the same way as before. What then was there to per-manently prevent the workers from producing on their own account, either individually or in small groups, and extricating themselves once more from wage dependency? They were now dependent on wages, but being in full possession of their specialist skills were in a position to produce complete, saleable products and were even able to make their own tools. Since they controlled the whole production process themselves, they didn't really need mediation by a third party (the capitalist) at all. It was precisely to escape this danger that capital became involved in the production process, and it was to suit capital that pro-duction was redesigned from top to bottom. Instead of each manual worker carrying out every operation necessary to manufacture a product as was the case before, the worker was now to undertake the execution of just one particular operation. The article being manufactured would now pass through the hands of each worker in turn.

With the introduction of this horizontal division of labour, a worker's knowl-edge and ability became specialized and related to a particular task. The uni-versal skill of being able to undertake many types of work, and with it the capability of producing complete marketable products, almost disappeared after a time. This then is how, in the absence of any other method of control, the principle of "Divide and Rule", which has been effective over and over again in other spheres, was used to ensure control by capital of paid labour in manufac-turing (Marglin 1977).

The same strategy that secured management control so effectively in manu-facturing, turned out to be a highly efficient economic measure as well. In the incessant pursuit of higher profits made imperative to stave off ruin through competition, individual capitalists discovered that this division of labour brought about increased output with the same labour and materials. This was on top of the other advantages of manufacturing such as its spatially compact production process, better co-ordination, supervision and control and the reduced call on transport. This splendid idea, the division of labour, was even more welcome when the old tactic of simply making wage-earners work longer hours to achieve higher production targets, was no longer an option because of the physical limits of labour power and somewhat later, because of legal barriers as well. Adam

Smith, who saw the division of labour as ideal for social wellbeing, summarized the economic advantages as follows:

This great increase of the quantity of work, which, in consequence of the division of labour, the same number of people are capable of performing, is owing to three different circumstances; first, to the increase of dexterity in every particular workman; secondly to the saving of time which is commonly lost in passing from one species of work to another; and lastly, to the invention of a great number of machines which facilitate and abridge labour and enable one man to do the work of many. (Smith 1776 p. 10)

Of course, his sense of reality also appreciated the drawbacks of the division of labour:

In the progress of the division of labour, the employment of the far greater part of those who live by labour, that is, of the great body of the people, comes to be confined to a few very simple operations, frequently one or two. But the understandings of the greater part of men are necessarily formed by their ordinary employments. The man whose whole life is spent in performing a few simple operations, of which the effects are perhaps always the same, or every nearly the same, has no occasion to exert his understanding or to exercise his invention in finding out expedients for removing difficulties which never occur. He naturally loses therefore, the habit of such exertion, and generally becomes as stupid and ignorant as it is possible for a human creature to become. (ibid. Vol. 2, pp. 263f)

Nowadays we know to our cost, that he correctly identified the advantages and disadvantages of the system. Nevertheless, as Charles Babbage later demonstrated to him using a meticulous calculation (which, as in Smith's case was taken from pin manufacture), he had overlooked the most important economic advantage of all, namely:

that the master manufacturer, by dividing the work to be executed into different processes, each requiring different degrees of skill and force, can purchase exactly that precise quantity of both which is necessary for each process; whereas, if the whole work were executed by one workman, that person must possess sufficient skill to perform the most difficult, and sufficient strength to execute the most labourious, of the operations into which the art is divided. (Babbage 1835, p. 175)

We have here what is known as the "Babbage principle"; fundamental to the development of the division of labour in a capitalist environment, that of lowering the value of the workforce by purchasing precisely those attributes needed for carrying out individual tasks of varying degrees of difficulty. By virtue of the fact that this principle actually works, a polarized workforce is formed and it then demands the application of the Babbage principle itself. The reason it does this is because at this stage the labour market can offer only a few jobs to highly skilled workers among a workforce which have few skills if any. "In this way all work processes obtain a structure, at one end of which are concentrated those whose time is infinitely valuable, and at the other, those whose time has practically no value at all" (Braverman 1977, p. 72).

We refer to this polarization of abilities nowadays as well of course, and I will return to it later. In the minds of many managers however, this occurrence is completely misrepresented. They perceive that the "lack of qualified employees" requires more efficient use of the ones they have and that it is worthwhile "retaining rare abilities". They reckon that most workers are "overtaxed" when faced with more demanding tasks.

But Babbage did not stop there. He immediately went on to consider the division of intellectual work, and in an age when almost all work was carried out through a unity of hand and brain. Purely intellectual work, which made up only a minute proportion of the total work performed in society, was seen to be particularly difficult. This must seem almost visionary. "We have already mentioned what may, perhaps, appear paradoxical to some of our readers: that the division of labour can be applied with equal success to mental as to mechanical operations, and that it ensures in both the same economy of time" (Babbage 1835, p. 191).

Babbage demonstrates this, using as an example the story of a certain Prony who, during the course of the French Revolution, had received the order to calculate logarithmic and trigonometrical tables in the newly introduced decimal system. He quickly realized that if he used conventional procedures he would not succeed in his own lifetime. It is alleged that by chance, he came across the passages on the division of labour in the newly published book by Adam Smith and this inspired him to use the following procedure.

He had the whole calculation carried out twice in succession so that errors would quite probably be corrected. To do this, he divided the workers up into three teams. The first team consisted of some of the leading French mathematicians. Their task was to develop the most suitable method of calculation. The second team was made up of eight people with good mathematical knowledge. They had to convert the method into calculation formulae and oversee the implementation of the process watching out for errors. The third team consisted of 60 to 80 people who had almost no mathematical knowledge. They did nothing but simple addition and subtraction and recorded their answers according to routines which had been prepared in the calculation process (ibid. pp. 191ff).

In Germany during the Second World War, the flight paths of the V2 rockets were calculated in the same way using over 1000 semi-skilled, mostly illiterate, foreign workers. They had the assistance of four species of mechanical calculating machines connected in three parallel sets so that it would be highly likely, not only that errors would be spotted but also that they would be corrected. The German computer pioneer, Konrad Zuse, used this method for calculating complicated functions in his development of computers.

2.1.2 Productivity Increased: Skills Limited

The radical social changeover to manufacturing and the division of labour in the factory also revolutionized production technology; but of course in quite different ways from those the protagonists of orthodox "technological progress" would have us believe. Nothing could demonstrate this better than a comparison between two completely contradictory designs for the mechanization of spinning: Hargreaves' "spinning-jenny" patented in 1770, and Arkwright's "chain-chair" patented in 1769 and sometimes called the "water-frame" because it was driven by a water mill.

The basic design of the spinning-jenny consisted of two parallel bars running along the spinner's line of sight, and a free moving carriage able to roll along these bars (Fig. 1). There was a strip-shaped clamp on the carriage, divided up according to the number of spindles (16 in the patented version, 20/30 later on), and the slubs could pass freely through this clamp. At a certain point in the spinning cycle, the carriage could be held fast with the clamp. A wheel with a hand-crank was mounted vertically on the right-hand bar, allowing the spindles to be driven by means of pull-ropes. In addition, an adjusting wire could be operated from the carriage to control the unwinding process.

Fig. 1. Basic construction of the spinning-jenny (here in a post-1770 version): 1 frame, 2 carriage, 3 spindles, 4 hand-wheel, 5 bobbin, 6 spools and spool-carrier, 7 drop wire, 8 work position (from Kuby 1980).

The essence of spinning with the jenny was that the carriage, controlled with the left hand, and the wheel, rotated with the right hand were operated together in the spinning cycle. The individual phases of the cycle could be co-ordinated with precision. Starting off as far forward as possible, and with the clamp open, the carriage was drawn back along the length of the thread to be spun. Then, the clamp was closed and the spindles set rotating at the same instant whilst the carriage was moved further in the same direction. During this phase, the thread was both twisted and stretched, and sometimes just twisted when the carriage was brought to a halt. Turning the hand-cranked wheel briefly in the reverse direction loosened the spun thread on the reel and finally, as the carriage moved into reverse it was tightened and ready for rolling up. For this to be done, the carriage was pushed forward again at the same speed as the thread was wound on to the wheel, while the hand-wheel, cranked in the original direction, was used to drive the spindles. During this process, the point at which the thread ran on to the thread cone was controlled by the adjustment wire.

This description shows that the whole course of the spinning process was regulated by the worker himself and that, at every instant he determined the force used, and the duration and speed of each operation (Kuby 1980, pp. 52f).

The same task was also performed by Arkwright's chain-chair which functioned according to completely different principles. Instead of spinning the thread in phases of intermittent hand movements, Arkwright's machine did it continuously and was mechanically driven by water mills initially, and then by steam engines. At the start of the spinning process, the slub passed through three rollers rotating with increasing speed which caused them to stretch. Then the thread ran on to a capstan which by turning, both twisted it and rolled it up. Thus, the basic operations of spinning, stretching, twisting and winding up, were combined by the mechanism into one continuous and automatic process from which the workers were excluded. Their sole duty now was simply to load and unload the machine with material and to deal with broken threads.

Both these inventive efforts were spurred on as a direct result of the enormous and constantly growing need for yarn which demanded increased productivity. The spinning-jenny was the final response of the age of craftsmanship to this problem. As a highly complex tool, it made the skilled work of spinning more productive. It was not a machine like the chain-chair which was conceived from the beginning as a replacement for skilled manual labour. The former was designed to increase output by responding to the intermittent movements of the hands controlled by the brain, without fundamentally altering them. In the latter however, the whole sequence of events was completely redesigned so as to form one continuous automatic process. This was done by dividing up the movements into their essential elements and then reconstituting them.

In contrast to the jenny, the chain-chair embodied two basic principles of all machine development in a capitalist environment, the first being the principle of decomposition and reconstitution, looked upon as the be-all and end-all of machine development; and the second being the principle of automation, accepted as the form of production most suitable for capital by rendering its processes quite independent of skilled human labour (Edwards 1981; Giedion 1982).

The main characteristic of the capitalist employer is his independence; for he merely purchases all the ingredients of production on the market (workforce, machines, materials, knowledge etc.) and has them work together in the production process in the correct functional manner without laying a hand on it himself. Consequently, it is not by chance but of necessity that manufacturing spawns a mechanistic philosophy. "As long as functional independence has not yet become a reality through machine technology, it rules merely in the minds of the capitalist class, but it disappears from these minds as quickly as it assumes a concrete form outside them" (Sohn-Rethel 1973, p. 154).

The further development of mule spinning confirms this fact too. Initially, Hargreaves' complex tool was still used on the whole for this activity, by people working at home in traditional craft surroundings. Arkwright's machine though, was intended for factory installation from the outset, because it could only be driven in a factory. Right at the start, Arkwright's invention exhibited all the qualities useful to capitalist industrialization including automatic operation, consistently high output and ease of operation for women and children. In contrast to this, Hargreaves' invention had first of all to be adapted for use in factories.

In a development process which lasted 60 years and which challenged the initiative of dozens of inventors, the jenny's complex spinning process was gradually mechanized, and in time, this complex hand-tool was transformed into a machine. The first important step was taken by Crompton in 1779 with his "mule", which mechanized the stretching process and truly represented a point of convergence with Arkwright's continuously operating stretching mechanism. Only in the years 1825 to 1830, following a major strike of workers, did Roberts manage, at enormous cost, to mechanize the whole process of mule spinning (Kuby 1980).

It was worth taking a detailed look at the development of the spinning machine, if only for the fact that here, right at the start of mechanization, the basic features of capitalist industrialization can be seen against the background of society's driving forces.

A further important example is provided by the mechanical loom. This loom also mechanized complicated skilled work in a novel manner. In 1804 Jacquard developed a control device in order to facilitate the regulation of its elementary movements according to the pattern required, and hence to make it capable of

Fig. 2. Turner with mediaeval pole-lathe (from Kuby 1980).

Fig. 3. Craftsman at a lathe without a tool-slide (left), and skilled worker at a lathe with one (after a drawing by J. Nasmyth, from Kuby 1980).

performing many different operations. The device relied on the use of punch cards to control the raising and lowering of the warp threads. Due to the positioning of the holes in the cards a woven pattern was revealed. This punch card device is an early example of flexible machine control. It is also the forerunner of modern industrial machine control, which was later termed "numerical".

Very similar technological strides were made in the case of machine tools, particularly in motorized lathes; that is, insofar as there were hand operated forerunners in the field. The rocking pole-lathe (Fig. 2), whose intermittent drive was set in motion with the foot by means of a rocker attached to a rope, had been in use for centuries. In the hands of a skilled craftsman, the rope wrapping the workpiece was a complex tool. He used it to perform turning work from start to finish. His mind controlled the entire sequence, regulating the force and speed used and in particular making sure that the cutting tool movements were correctly synchronized when rotation of the workpiece was reversed.

The qualitative leap forward which transformed the tool into a machine and the skilled craftsman into a specialist worker, took place in about 1800 when Maudslay introduced the continuous drive and the tool-slide in which a cutting tool was firmly clamped, and which was driven by a lead-screw (Fig. 3). Here again, we see a process taking place independently, at least in part. The invention and successful use of all these machines could only come about by breaking up the manual work process into its recurring operations, converting them into elements of mechanical activity, and then assembling them into a formal sequence of events (the theoretical construction of a model).

Industrialization itself caused the problems in part, and the solutions provided further stimulus to action. Thus as we have seen, the compulsion to use machines in the course of the upturn in the English textiles industry required the use of drive mechanisms that did not have to rely on water power. So this industry, along with the mining industry where there was a continual need for increased pumping power, created a need for improved steam engines. Watt's double-action steam engine of 1784, with its flywheel, was a good example. The construction of these machines, and later that of machine tools themselves, demanded substantially more accurate methods of production, and this led to new designs for machine tools (for example, Wilkinson's mechanically driven cylindrical drill (1775), or Maudslay's motorized lathe required principally for producing lead-screws). The industrial revolution was thus borne along by machines for doing work and not by machines for producing power.

These radical changes reveal how interacting social forces alter so fundamentally and at an unpredictable rate, when machines based on the division of labour are developed and used in the factory. When these radical changes take effect, everything else is made to appear as autonomous "technological progress". It is the more or less far-reaching autonomy of the machine, which the capitalists are constantly aiming to extend, that provides them with their desired high degree of independence from the wishes and abilities of human labour. This autonomy largely determines for labour, the nature of the work process and to that extent, their own control over it is removed. At the same time however, the use of machines is a proven means of increasing output and profits, so as long as the money spent on using the machines for production does not exceed the costs of the labour saved through their use.

Recently, a radical view of the machine has become common. This takes only its functions into consideration and removes it from the context in which it originated. A typical example of this technocratic, functional view is given by Reuleaux in his definition of the machine. "A machine is a combination of resistant bodies which are arranged in such a way that, with their assistance, the mechanical forces of nature can be compelled to perform work which is accompanied by certain precisely defined movements" (quoted by Braverman 1977, p. 144).

Apart from the fact that this definition still relies totally on the mechanistic model for explaining the world and is manifestly inadequate in the case of machines that process symbols, it disowns any connection with the work process. That is, it rejects the interrelation between the worker with a goal in mind, the working materials, the machine itself, the article being worked and the social interests which shape this process.

In contrast, the fact that machines emerge in a social context, where they are developed and used as a means of work and where that work leaves its mark on them, did not escape the keen eyes of Babbage, a contemporary active observer:

When each process, by which an article is produced, is the sole occupation of one individual, his whole attention being devoted to a very limited and simple operation, improvements in the form of his tools, or in the mode of using them, are much more likely to occur to his mind, than if it were distracted by a greater variety of circumstances. Such an improvement in the tool is generally the first step towards a machine. When each process has been reduced to the use of some simple tool, the union of all these tools, actuated by one moving power, constitutes a machine. (Babbage 1835, pp. 173f)

Unlike the technocratic, functionalist observers who generally viewed the machine as an instrument for man's happiness that would free him from arduous work, Babbage also had clear ideas about the consequences and drawbacks of using machines. For in a capitalist environment, the purpose of the machine apart from its technological function of increasing productivity, is still to deprive the worker of control over his work. "One great advantage which we may derive from machinery is from the check which it affords against the inattention, the idleness, or the dishonesty of human agents" (ibid. p. 54).

It was also Babbage who laid the foundations for the mechanization of intellectual work on which all computer design is based, even today (apart, that is, from the most recent configurations for parallel processing). The theoretical analysis of the most varied intellectual processes, particularly the wide ranging and complex calculations for which he was trained as a mathematician, made him realize that they too could be reduced to a few basic and recurring arithmetical and logical operations performed in an appropriate sequence. Out of this arose his idea of a program-controlled universal computer. Once again the principle of breaking down and reassembling was applied.

The computer's mechanism performed these continuously recurring basic arithmetical operations. Input data, provisional results and output data could be stored in its memory in an orderly fashion and its control mechanism arranged the whole process according to instructions contained in a program which was input from outside as a series of symbols. The vast range of programs that could be processed was what determined the universality of the machine.

Babbage achieved technical success in the form of his "difference engine", which was less complicated in design, but still able to carry out mechanically, the enormously complex and error prone calculation of logarithmic and trigonometrical tables. His development of a more versatile computer, the "analytical engine", finally foundered because of the inaccuracy of the mechanical parts and the shortcomings in their manufacture; it was not due to the inadequacy of his design. For example, the memory alone needed 50 000 extremely accurate gearwheels to represent the figures. It wasn't until 100 years later that this great blueprint for a universal computing machine was turned into a reality, when relay techniques and later electronics, created a new technological base to replace mechanical components. Now the pressure was on to rationalize intellectual work, for a point had been reached where enormous expenditure on development carried with it, the promise of reward (Brödner et al. 1981).

This brief glimpse at the development and use of machines is quite adequate for our purposes. We have deliberately restricted ourselves to a few examples which throw light on the background in which they appeared. Rather than pursue as thoroughly as possible, the many lines of development associated with different types of machine, it is preferable to give an idea of the historical context in which they emerged; for their designs were determined by the specific social driving forces of capitalist production. Recognition of this fact is lacking in most historical treatments of the development of computers; for although these writings are rich in museum pieces and in meticulous descriptions of how they worked, they are lacking in historical content. They are like a pile of fragments that no one has tried to reassemble into the vase they came from.

2.1.3 Taylorism: Dividing Hand and Brain

When machines first entered the world of capitalist production, they needed to be perfected in order to fulfil their two main purposes. One was to increase the creation of surplus value, and the other to guarantee the autonomy of the production process by forcing the laws of commerce on to stubborn, human labour. Apart from a few fundamentally new inventions such as the milling machine (Whitney, 1818) and Brown's universal milling machine (1862), the existing machines were progressively improved in order to achieve three main aims. These were to raise output, to increase precision and thus facilitate the production of interchangeable parts and finally, to enable operations to proceed more independently.

Quite early on, development branched into increasingly specialized machines for the mass production of standard parts (such as screws), consumer goods and armaments (especially in the USA) and machines of a more universal design for the production of a whole variety of capital goods. These machines shaped industrialization in England and in the German Reich.

In 1870, Ludwig Loewe, an enthusiastic pilgrim to the USA, which at that time was the Mecca of production technology, caught the mood of capitalism very well when, on his return he announced that the use of "automatic machines" would make it possible "to become independent of the goodwill and technical capabilities of any single worker" (quoted by Boberg et al. 1984, p. 324). This aim was only ever achieved to a limited extent and then only in particular operations and to greatly differing degrees. It happened with spinning and weaving machines and in the production of screws (e.g. Carver's multi-spindle automatic lathe of 1879) but to a much lesser extent in the case of universal machine tools for short batch production. This is not to say that a sudden lack of interest in restructuring on the part of capital set limits on the use of machines, but rather that the theoretical model was inadequate. Not many operations, not even the simple ones occurring over and over again in mass production, are so open to design analysis and theoretical description that they can be made automatic. For production, however much it is organized on the basis of the division of labour, will always contain some constantly recurring operations which have to be performed by a skilled hand controlled by the brain.

The workers were not particularly inclined to subject themselves peacefully to the discipline of the factory. Their entirely rural backgrounds based on craftsmanship, made them unwilling to accede to the demands of management for a fixed rate of working and for stamina, diligence, punctuality and obedience. Factories thus experienced a constant battle to maintain the order and discipline they required. There was continual compromise between resistance and attempts to adapt, such that the rules in some factories were exactly like parade-ground regulations. Even piece-work, which was introduced early on and was supposed to provide workers with a material interest in higher output, could in effect change none of this. From the beginning, the workers recognized this as the employers' Trojan horse for demanding the greatest amount of work possible, but instead of taking control of their own labour power, they reacted by slowing down output. There was at first no way of stopping this "dawdling", since the workers possessed practical knowledge which management largely lacked. It meant that fundamentally new forms of control had to be found to break the stalemate (Pollard 1973; Thompson 1973; Braverman 1977).

Even though the efforts to objectivize production knowledge through machines did not entirely succeed, management was at least able to make the attempt at transforming part of the workers' vast experiential knowledge into abstract planning knowledge (expropriation of the knowledge of production). No less a figure than Krupp, one of the leading lights of German industrial capitalism, formulated this idea with commendable clarity in 1874:

What I wish to strive for is a situation where nothing is dependent on the life or existence of a particular person; for no person's knowledge or ability to be allowed to slip away; for nothing of crucial importance to happen or have happened, which is not known to central management or is not happening with the foreknowledge and approval of the management and for it to be possible for the factory's past as well as its probable future to be studied and clearly viewed in the manager's office, without recourse to any human being (quoted by Kocka 1969, p. 277).

The abstract model of the production process really did open up a new dimension for management, in gaining the upper hand and exercising control. The model was constructed by meticulously analysing the process and then using this analysis to determine the best possible way of working, irrespective of the wishes and abilities of the workforce. The model was then imposed on the workforce by means of detailed rules and regulations, and checks were carried out to make sure they were being adhered to.

The investigation and analysis of the production process and subsequently its appropriation by management on an academic level, were indicators of a transition from the old barracks-style disciplining of the workers to the planning and control of their work. The way this could be done was demonstrated by Taylor in the Midvale Steelworks from 1880 onwards, using the complex process of metal cutting. Over a period of 26 years he performed well over 50 000 metal-cutting experiments, during which time he machined 400 tons of iron and steel into chips. He emphasized that:

the motive power which kept these experiments going through many years, and which supplied the money and the opportunity for their accomplishment, was not an abstract search after scientific knowledge, but was the very practical fact that we lacked the exact information which was needed every day, in order to help our machinists to do their work in the best way and in the quickest time. (Taylor 1911, p. 106)

In order to answer the apparently simple question of what the best cutting parameters (cutting speed, rate of feed and depth of cut) were, he studied no less than twelve variables including the quality of the cutting tool steel and the material to be worked; the form of the tool's cutting edge and its useful life; the chip thickness; the amount of coolant and so on. The complexity arose from the need for eleven of the variables to be kept constant each time so that the effect of the twelfth could be studied. As a means of determining the best cutting parameters in the simplest way, given the complicated relationships between them, he finally came up with his "slide-rule" for metal working workshops (ibid. pp. 107ff).

Filled with faith in the universal success of this procedural method, he proclaimed it in a general form as the Principle of Scientific Management. Making sharply critical remarks, he distanced himself from the old factory methods which he mocked as "enticement", whose success depended almost entirely on how successfully one could induce "each worker to use his best endeavour, his hardest work, all his traditional knowledge, his skill, his ingenuity and his goodwill, in a word, his 'initiative', so as to yield the largest possible return to his employer" (ibid. p. 32).

Instead of this, it is up to the factory owner, in accordance with Taylor's first principle, to take on "the burden of gathering together all of the traditional knowledge which in the past had been possessed by the workman and then of classifying, tabulating and reducing this knowledge to rules, laws and formulae.

They develop a science for each element of a man's work, which replaces the old rule-of-thumb method" (ibid. p. 36). In this way, management acquires objective knowledge of production, independent of the experiences of the human workforce and uses it to sever the connection between production and skills.

Taylor's second principle is directly linked with this:

The development of a science involves the establishment of many rules, laws and formulae which replace the judgement of the individual workman and which can be effectively used only after having been systematically recorded, indexed etc. The practical use of scientific data also calls for a room in which to keep the books, records, etc., and a desk for the planner to work at. Thus all of the planning which under the old system was done by the workman, as a result of his personal experience, must of necessity under the new system be done by the management in accordance with the laws of the science. It is also clear that in most cases one type of man is needed to plan ahead and an entirely different type to execute the work. (ibid. pp. 37f)

This principle separates intellectual work from manual work and its planning from its implementation. The manual worker is conscious of being limited to purely operative tasks, and the objectivized knowledge of production which has been taken from him, now confronts him as an alien force in the work schedule, which it is from now on his duty to follow meticulously. As a follow-on from horizontal division, vertical division of labour has now become established as well.

The most significant fundamental feature of the new system is embodied in Taylor's third principle, the "target idea":

The work of every workman is fully planned out by the management at least one day in advance, and each man receives in most cases, complete written instructions, describing in detail the task which he is to accomplish, as well as the means to be used in doing the work. This task specifies not only what is to be done but how it is to be done and the exact time allowed for doing it. (ibid. p. 39)

The monopoly on objectivized knowledge is thus used to specify each step of the work process and to control it when it is implemented.

Taylor's system had now finally provided the means of control, and thanks to the Babbage principle which we have already met, it was possible, in line with the second rule, to employ a distinctly less well qualified workforce in its implementation. The dual advantage of not only securing control at a new level through the new structure, but also paying less for the essential workforce, gave the Taylor system the necessary drive to complete its victorious progress through the companies and to revolutionize the production process. Right up to the present day, its principles have determined capitalist restructuring plans in manufacturing, in spite of their appearance under many other guises.

It is true to say that from time to time, the introduction of the Taylor system gave rise to strong opposition within the ranks of skilled workers, but in the end, they could not stem its advance. As a rule, it was possible to lessen, or even overcome, the initial resistance of skilled workers and their union representatives by promising them higher wages, and their obstinacy could often be tamed by holding out the prospect of a better life (Neubauer 1981). Occasionally however, drastic measures were taken as for example, during the strike of skilled workers in Berlin in 1906. In the course of the strike, the electrical company Siemens sacked all 4000 skilled workers and with little hesitation, replaced them with unskilled workers (Boberg et al. 1984, p. 316).

Entirely different circumstances resulted in a setback for the Taylor system. Often, production processes proved to be so complicated that management could not work out any suitable plan. So, as before, they were dependent on the co-operation of skilled workers, though to a lesser extent. This was because some processes were dependent on events and market forces that could not be calculated or controlled and were thus unsuited to theoretical modelling. Taylor's principles could still be applied through the techniques of mass production. Here both the variety and complexity of the production process could be greatly restricted in comparison with workshop production; just as, by meticulous planning and with even the smallest improvements, extensive rationalization could be achieved.

Since mass production in expanding markets has an innate tendency to deprive customers of choice and to encourage them to buy what makes sense from a production engineering standpoint, there is a reduction in the customer's input on product design and also on production. Of course, any input can itself be influenced to proceed in predetermined ways.

Taylor's ideas penetrated the most, and spread the fastest in the USA where a huge and expanding domestic market caused mass production to flourish more quickly and effectively, but they were also adopted by Ludwig Loewe, Siemens, AEG and other manufacturers of mass products, whose factory managers copied the newest manufacturing methods.

Now, the more these ideas succeed economically, in the ever expanding market, the gloomier and more threatening the other side of Taylorized mass production becomes. The production process as a whole now specializes in the few products that are mass produced. Everything is calculated and planned to reach a definite level of output. Its individual processes and machines are divided up into the smallest elements possible. They are specialized in their particular function and their capacities and configurations are co-ordinated and synchronized to such an extent that no delays or hold-ups occur. If individual parts of the process were to affect the whole in a disporportionate way, the capital equipment would be functioning at a substantially lower rate of profitability. To change the level of output would require radical structural changes.

A production process of this type has sacrificed its ability to adapt to variations in demand. It reacts to reduced demand in the opposite way to that required by the market since its unit costs will increase. Consequently, if it cannot function with the higher throughput, it functions unprofitably; that is, unless there is a monopoly that can maintain higher prices when demand falls. The "quantum economy" of production conflicts with the "continual economy" of the market place (Sohn-Rethel 1973, pp. 179f).

There are however some situations in the labour market and sales market that are actually, quite decisive factors in advancing Taylorism. If there is a relatively long-term shortage of qualified workers at the same time as the market for the product is expanding, businesses will have no other choice but to Taylorize production as much as possible in order to be able to produce the goods with workers who are inexperienced in metalworking. There are plenty of examples of this which come to mind; in the USA throughout the industrial upturn, in France, in Nazi war production and even to some extent, in West Germany in the 50s and 60s. The key to this Taylorization is the availability of streams of refugee and immigrant workers at a time when skilled workers cannot be obtained.

The situation is different in short batch production, which characterized the German industrial upturn. In the capital-goods industry the production is tied to orders. Here, the client is in charge, and his wishes have much more influence on the products and on the production deadlines. Production becomes much more complex and varied and consequently, less easy to plan. Certain aspects of the process, in particular the sequence of operations, their implementation and duration, can in fact be planned in advance in accordance with Taylor's methods, but not the process as a whole. Even so, much is based on rule-of-thumb and estimation, and so many non-calculable and unpredictable events may arise, that management can only overcome them with ad hoc improvizations carried out by skilled workers.

In this way then, the application of Taylorism in this sector of work, always comes up against the limits of model building, and brings about the typical hybrid structure found everywhere. With its workshop organized around routine tasks, a foreman, centralized work planning, horizontal, and vertical division of labour and a high proportion of skilled personnel, it will already be a Taylorized factory; yet it will still be a mechanized workshop as well, in part.

In Germany, it has been possible to maintain a powerful section of permanent specialist workers in the industry. These act as a link with those parts of the production process which require human ingenuity. These workers, who are capable of being highly flexible, have capabilities which are of course, used to only a very limited extent.

2.2   The Computer and the Automation of "One-offs"

2.2.1   The Vain Hope: Unskilled Flexible Automation

The hybrid character of short batch production remained a constant source of irritation for management, not only because it left unrealized, the Tayloristic promise of total control, but also because the large numbers of qualified staff required in that type of manufacturing, drove up costs without leaving room for increased output. Thus the profitability of the capital used became endangered. At first, automation was out of the question for as we have said earlier, the process defied analysis and even more, it had to respond to demands for flexibility. It was later, during the Second World War, that storage and control technology developed for air defence, presented an opportunity for approaching this problem anew. Just as in the traditional stage of manufacturing in the spinning industry, two very different technological solutions were put forward simultaneously; both based on existing technology. Both solutions depended on flexible automation of metal cutting, which meant the automatic operation of machine tools without loss of versatility.

Of these two solutions, the "record–playback" one was developed by General Electric and Gisholt. The first part of the production run was produced by a qualified worker as normal, but all the motions and operating stages of the machine tool were recorded using measuring techniques, and these were recorded on magnetic tape. This tape, on which the data had been stored, was then used to control the production of all the remaining parts in the series. This splendid technological solution found little favour with potential users. General Electric offered it as an option for machine tool control, but it did not bring the resolution of the factory control issue any nearer. Just as before, the skilled workforce held the reins and determined the course and efficiency of the work. This was the response of the old style skill-based workshop to the inadequacy of the semi-Taylorized factory, and so it was soon rejected (Noble 1979).

The other solution, the numerical control (NC) of machine tools, was introduced at the Parsons factory in the production of rotor blades for helicopters. Shortly after that, it was taken up by MIT for further development. The information required for the control of the machine tool was "fed in" in the form of numerical data on a magnetic, or punched paper tape. The control mechanism co-ordinated the slide's movements, which were being monitored by measuring equipment. The basic principle was not new, indeed we have come across it already in the case of Jacquard's loom. In this case however, it was only possible through the use of new technology, because the complicated and extremely precise motions involved, meant that it had to be considerably refined.

Now that this principle had been applied in practice, the decisive step had been taken. The NC program containing the control data is no more than an algorithmic description of the working procedure which can be drawn up in the technical office. There is no need for skilled human labour. The geometrical data can be taken from a drawing, and the control data can be extracted from tables of cutting values drawn up during preparation for the work.

Organizing production in this way turned out to be very costly, but since it promised efficiency in producing complex geometric forms such as surface profiles and the profiles of turbine blades in aircraft, and since too, it was independent of the specialist skill of the individual worker, it gained massive support from the US Air Force from the outset. It was designed to undertake complicated milling using five numerically controlled axes. The creation of an NC program nevertheless very soon turned out to be something like a black hole where the only protection from its all-consuming suction was the programming language APT developed for that purpose. Using APT, the geometrical shape of the workpiece could be defined and from this, the sequence of movements of the tool could be calculated by the computer itself.

The concern of the US Air Force for a general, unified, and versatile NC programming method was at first, of material interest to only a few firms dealing in aircraft and machine tool manufacture. The Air Force itself supplied the specifications, controlled the standardization and in reality created the first "market" for NC technology (Noble 1979; Shaiken 1980).

The innovation was accepted only hesitatingly outside this military and industrial complex. It was used by larger firms, manufacturing machine tools and control mechanisms, but only for the lack of anything better. The first 100 machines were subsidized by the Air Force. Only when greatly simplified programming methods were available was there greater use of the system. For example, there was the EXAPT family of languages, which were subsets of APT based on process technology. Nothing changed though, in fundamental working methods. Strict separation of planning and implementation was the rule, and NC programming was always intended for planning the work, which would then be carried out by computers.

It seems that this Taylorist solution had finally provided a way of gaining independence from the knowledge and abilities of human labour, and of controlling the work process as efficiently as possible; even in short batch production. The level of control that NC technology made possible in the factory was the reason it found favour with management. It left the "record–playback" process with no future. Even when at first, the economics of the method were extremely questionable, the promise of control helped it to break through. There were some considerable economic advantages in it later, such as the saving in equipment, the reduction of preparation time and improvements in quality (Brödner and Hamke 1969, 1970).

The dream of a factory with NC machinery and no skilled workforce was of course, soon shattered. A rude awakening revealed that with high requirements for performance together with demands for precision, no process could be carried out in the completely predefined way originally envisaged. Deformation of the machine tool, wear of the tool, stretching when the workpieces were under stress and other such unforeseen technical influences, caused the actual process to continually deviate from its programmed sequence and this necessitated expert correction. In addition, high capital equipment had to be used together with the co-ordination operations that went with it. Work to be carried out on the programs, tools and workpieces organized on the basis of the division of labour, had to be combined and co-ordinated and all this made it seem advisable not to dispense with qualified, skilled workers after all. Every attempt to replace these workers with semi-skilled labour ultimately failed.

Skilled workers now found themselves in another hostile situation. Owing to the programmed working sequence, their actual specialist qualifications and intellectual capabilities were no longer in great demand. It was rather their talent for organization and their general reliability that were needed, but these attributes were called upon only rarely. Instead, these talented workers were given simple, non-specific supervisory tasks and the mundane job of clamping and releasing the workpiece. They were left with practically no influence over the way the work progressed, and frequently suffered from stress due to lack of a challenge. They were constantly in conflict with the programmers in the technical office, who were writing mostly erroneous programs, without reference to the skilled workers on whose knowledge and experience they were dependent. This was a cause of continuing instability within the workforce. The personal friction it engendered led to staffing losses, for a constant small-scale war was being fought, particularly over wage structures and grading.

The expansion of this technology was therefore restricted to a relatively small number of pioneering companies with sophisticated forms of work preparation. It bypassed the majority of small and medium-sized companies (Noble 1979; Kern and Schumann 1984).

Yet again, the ambivalent nature of a situation in which a technology based on Taylorism demanded qualified workers needed to be clarified. The renewed CNC technology (computer numerical control), which replaced hard-wired control mechanisms, with computers controlled by software, gave NC technology a renewed boost. In particular, the computing power contained in the control mechanism made it possible to design programs more efficiently and on the spot. Hence, NC technology came to be used in small and medium-sized firms from which it had remained largely excluded up till now.

This was clearly the case from the middle of the 1970s when there were rocketing sales of CNC machines. In 1974, only 4000 NC machines were in use in West Germany, but by the middle of 1980 this figure had risen to 25 000 (Rempp and Lay 1981; Leibinger 1983). Very soon, a fair selection of designs was available ranging from pre-programmed machines, through the kind of machine whose programs could be modified on the spot by a specialist, to ones where all the programming was done by skilled workers on the shop floor. The decision on which kind of machine to use depended overwhelmingly on the employment policy of the firm itself, and on the organization and qualifications of its workforce. The decision was not made simply on technological considerations (Rempp and Lay 1981; Sorge et al. 1982; Kern and Schumann 1984).

Once again, the work process had proven to be far less suited to analysis, theoretical investigation and consequently to prior planning, than had been thought. The enormous extent to which deviations from the theoretical model can be coped with in the actual process, and its consequences for economics and the politics of work, will be discussed later, in Chapter 4.

Bearing in mind the sequence of events described here, it is not surprising, that in the many firms which relied on a specialized workforce, NC technology, could only make a breakthrough by introducing efficient methods of workshop programming, and that workshop programming is now moving forward, even in large firms with a pronounced vertical division of labour.

Despite the intervention of management, skilled labour was able to assert itself in the workshop. It has to be said then, that the original intention of management, which was to gain independence from the capabilities and knowledge of human labour by means of NC technology, can be judged with confidence, to be a technological and organizational flop.

2.2.2   Automating the Office

Under the pressure of competition, manufacturers of capital goods are forced continually to redesign their products, increasing quality and extending versatility. This is the reverse of what happens to mass-produced consumer goods, where the price is the decisive factor in remaining competitive. As far as capital goods are concerned, it is their quality and versatility that determine whether they will be purchased. A product designed to suit a customer's special requirements is the only one that will find a buyer in the market for capital goods, and this holds true even for those customers attempting to reduce their product range to a few basic items.

These aspects of competition in capital goods make profitable production exceedingly difficult, and the situation is intensifying as new international forms of the division of labour are being set up.

As a result of this market demand and in combination with the vertical division of labour within the factory, increasingly specialized functions develop in the area of technological and scientific, intellectual work. The range of this activity widens out enormously in such areas as tendering for orders, the design and construction of products and components, and in production planning. It is widened still further by NC programming. The number of tasks in dealing with orders are increased as well. They arise in the areas of purchasing, quantity surveying, and in the control and supervision of production. This phenomenon results in an enormous increase in the number of personnel employed in such tasks, with no proportional increase in the rest of the workforce. Tables 3 to 5 on pp. 92–93 illustrate this situation in mechanical engineering.

However, the constraints on capital cannot allow the cost of this necessary, but unproductive intellectual work to go on increasing for ever. The greater the range of products and the more complicated the production, the more extensive the aforementioned tasks become, and so the stronger the pressure for their rationalization. The methods used to increase output and reduce costs are totally conventional at first, and apply precisely the same principle used earlier in the rationalization of manual work, that is, the division of labour (the Babbage principle), involving the use of simple tools such as formulae, tables, mechanical adding machines and typewriters.

The Babbage ideal of a programmable universal computer became a reality for the first time in the 1940s when the Second World War helped to bring it about. As expected, it rapidly became obvious that here was the perfect and universal means for rationalizing intellectual work. Here too, just as in the sphere of manual rationalization, the fundamental prerequisite is to build a conceptual model of the work first, then analyse it and finally reconstitute it as a sequence of operations which will lead to a solution of the problem. Such a sequence of instructions is called an "algorithm". It generates a series of output data from a series of input data, and is the essential component of every program. The data are symbols representing information that can be transmitted on the basis of agreed conventions. They are abstract representations of the subject under consideration (the workpiece in the case of manual work), or of some of its properties. As soon as an intellectual work process has been theoretically investigated and analysed to a point where it can be described in the form of data and algorithms, it can be automated by means of a computer (Brödner et al. 1981; Bednarz et al. 1984).

This basic requirement for rationalization points to the primary and most obvious intellectual areas for computerization in the scientific and technological work of the factory. They are to be found wherever intellectual work can be expressed in algorithmic form and where pressure for rationalization is particularly high. In product design, calculation tasks are the first to be automated since they already exist in algorithmic form anyway, in particular, ones concerned with the static, dynamic and thermal strengths of materials. Workpiece drawings too are drawn up automatically in the early stages using computer-controlled drafting machines.

We have already touched on the intellectual work involved in the drawing up of NC programs. It is extremely complex but relatively easy to express in algorithmic form. Using the language APT, a method was designed which enabled the movements of the tool to be derived automatically from the geometrical configuration of the workpiece. The development that followed was at first restricted to these kinds of tasks by extending them, and by improving the efficiency of the programs.

There were early areas of use for the computer related to the processing of orders. In the metal industry it was used for stock control, for generating parts lists and for planning production capacity. These could be computerized with relative ease and with great economic benefit. Although the algorithms are all simple ones, the quantity of data which has to be stored, administrated and frequently amended is often huge. These early computers were used as centralized data-processing installations operated in batch mode. The elaborate accessing procedures placed limits on their expansion. Consequently, the pressure for rationalization built up before this problem could be sufficiently resolved.

2.3   Today's Factory: Unresolved Problems

2.3.1   Capital Intensity: Ever Growing

The increasing use of machines in production and the displacement of human labour from the production process have left their mark on the structure of capital. The proportion of the total industrial capital invested in buildings, machines (hardware and software) and means of production is increasing all the time. Taking capital intensity (the gross fixed assets per worker) as an indicator, we can quantify the development of the metalworking industry in the Federal Republic of Germany (see Fig. 4). During the period 1970 to 1983, capital intensity increased by almost three quarters.

There is no end in sight for this trend, nor even a clear reduction in it from the technocentric viewpoint. On the contrary, an uninterrupted growth in capital intensity is to be expected as we will show later in some detail. In spite of the fact that further technological change will involve computers whose central processing units are halving in cost every two years in terms of the cost/performance ratio, this is in no way the position with respect to the software, which nowadays accounts for over 80% of computer costs. As the range of functions increases, higher investment is required and capital intensity expands still further. This trend creates enormous efficiency problems in the production economy, especially since the machines can only be actually used for a comparatively short time compared with the theoretical maximum (see Fig. 4). The more capital intensive production becomes, the greater the pressure to overcome these problems by technological and organizational means.

2.3.2   The Vicious Circle of Stocks and Throughput Times

The vertical division of labour necessitates long preparation times for work planning, and the performance of routine tasks in a workshop production environment is responsible for long and often costly throughput times. On average, a production order is being worked on for only 5% to 10% of its total throughput (preparation and processing) time. All the rest of the throughput time is taken up by transfer from machine to machine and more importantly by storage (see Fig. 5). Not only does this situation create long delivery times and unreliable delivery deadlines, but it also adversely affects the cost structure, when stocks are by far the largest factor.

An analysis of the accounts of 95 mechanical engineering joint-stock companies in 1980 revealed that on average stocks accounted for 41.3% of the assets (ten years earlier it was only 39.4%; see Fig. 5, VDMA 1982). When it is considered that stocks of DM 1 million entail costs of about DM 200000 per year, it will be appreciated what enormous potential there is for reorganization, even if only a small proportion of the stock could be eliminated.

Moreover, lead times and stocks have a tendency to go on increasing in accordance with the following sequence of events. Owing to long storage times, management attempts to get work on production orders started early on so that there is no risk of missing the deadline. However, because of the large range of throughput times, it is still not certain that all orders will be ready on time. The

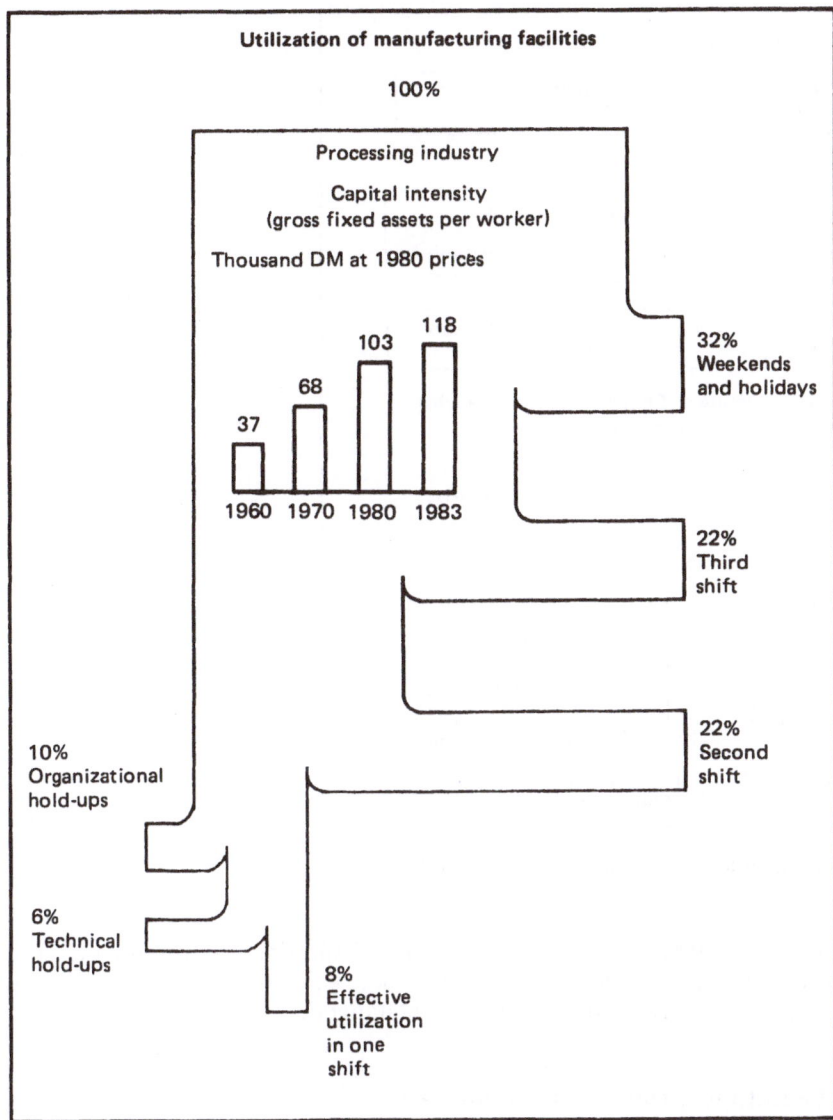

Fig. 4. Utilization of capital-intensive production facilities. Source: *Statistisches Jahrbuch 1986.*

more that deadlines are actually exceeded, the greater the inclination on the part of management to begin dealing with the orders even earlier. The problematic outcome of all this is even longer throughput times and even larger stocks in the workshop, without any reduction in the risk of deadline dates being overrun. Breaking out of this vicious circle requires a suitable strategy for releasing orders to production. The order release system based on workload, is aimed at carrying out production orders only when they have become urgent and when there is sufficient free capacity available for the work to be completed (Kettner and

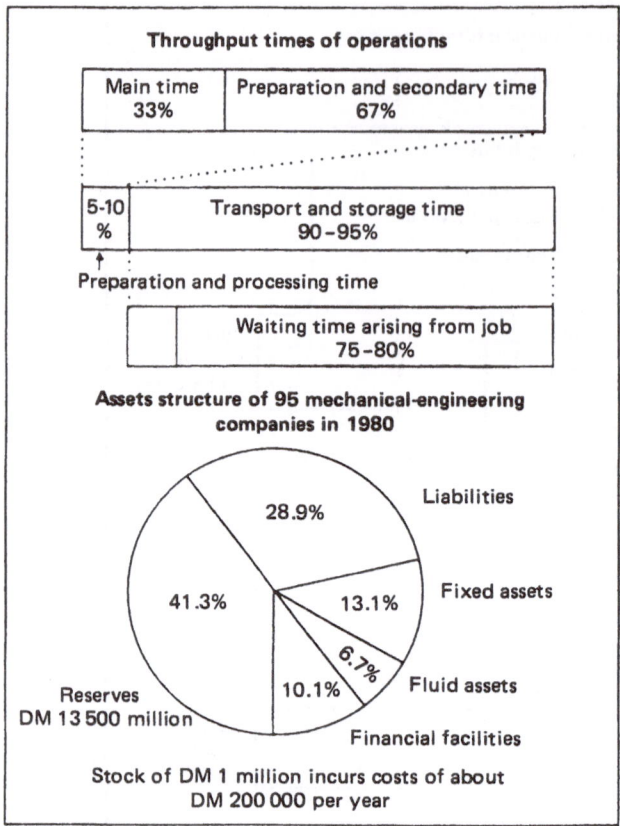

Fig. 5. Throughput times and stocks. Source: VDMA.

Bechte 1981). Of course, this strategy can only substantially reduce the effect of the vicious circle; it cannot get rid of it altogether, since the real cause of it, the principle of job shop production, remains unchanged.

2.3.3 The Inappropriate Work Structure

The vertical division of labour has compounded undesirable work structures in factories, simultaneously pushing up labour costs and restricting productivity. According to a study of 5500 mechanical-engineering firms employing an average of 200 workers, which was carried out by the VDMA in 1980, 59.3% of employees on average, work on indirect tasks such as planning, work preparation, material procurement, quality control and the like, and only 40.7% are directly involved in producing the product itself (Moll 1983). This structure reveals yet again the contradictory nature of small-scale production, customer led, in which planning and implementation have been separated, but which is nevertheless dependent on qualified, skilled labour to a very large extent.

The structure drives up costs. However, skilled workers have to be employed in order to ensure the smooth running of the complicated production process and the efficient operation of the expensive NC machines, but as we have seen, the abilities of these workers are only occasionally and partially used. On top of this, an excess of technological white-collar work has been created in the technical office for the planning and control of production, and this demands essentially the same abilities that the skilled workers have to offer anyway. It is not unusual for skilled workers to be "promoted" to work in the technical office.

Another effect of this is in the excessive production overheads incurred by companies with a pronounced vertical division of labour. According to a study carried out for VDW, the German machine-tool manufacturers' association, by Roland Berger and Partners, the proportion of production workers employed in "overheads" posts is two thirds higher in firms employing from 251 to 1000 workers than in firms employing from 51 to 100 workers. The vertical division of labour is evidently an important cause of these negative economies of scale.

Based on the availability of the skills required, there is of course, no objective necessity for a structure of this type. The fact that firms using the division of labour to a much smaller extent can produce efficiently, is evident from larger "craft" businesses which are certainly comparable with small and medium-sized "industrial" businesses when one takes into account the number of employees, the range of products and the machinery used. A study by the Bavarian trade corporation, covering 700 firms employing an average of 150 people, has shown that for every 100 workers directly involved in production there, only 86 are indirectly involved, compared to the 144 in mechanical engineering (Moll 1983). This comparison shows how inappropriate the structures are when considered on the basis of skill requirements.

Over the last three decades, there has in addition, been an overall decline in productive forces, indicated by a lowering of the rate of productivity increases. For example, the five-year moving average of the rate of change of the total economic productivity in the Federal Republic has steadily decreased from 7% in 1952 to 2.5% in 1980. In the metalworking industry, the average annual growth in productivity was 6.1% between 1971 and 1973, whereas between 1974 and 1980 it was only 3.9%.

The reason for this is not (say) that relatively fewer machines were used, because between 1960 and 1980 the rate of investment fluctuated between 22.0% and 26.5%, with no reverse trend (IFO 1982). Consequently, the reasons are more to do with the implementation of the division of labour (Sorge 1985). Rigid vertical division of labour in particular, which had already pushed up costs, now provides us with the reason why the abilities of skilled workers are restricted until they gradually wither away, and why too the indirect areas of planning and control, expand so much that they are regarded as greatly overstaffed. The inappropriate work structure may also help to explain the decline in the rate of productivity increases. The fact that industrial productivity depends to a large extent on the skills of the workforce has been corroborated empirically. In a comparison between British and German production companies, the higher productivity in Germany could be clearly related to the higher and more appropriate skills (Daly et al. 1985).

2.3.4   The Inappropriate Attitude to Costs and Benefits

Investors require an economic assessment to be made of the costs and benefits of whatever new production system they will assist in creating. Understandably, the well-tried procedures for this, such as calculating the internal interest rate or the return on investment, prove to be of limited value for the present assessment. These methods were more than adequate at the stage when it was simply the case of dealing with the purchase of a single machine, which had no effect on the structure of the production process, and was required for a particular job. When investment becomes linked with structural changes, such as those which occur when NC machines are introduced, the relevance of the standard accounting procedures is called into question.

The conventional cost/benefit calculations deal with production equipment in isolation, making use of data that at present only makes sense in financial terms, and which can in fact, only be dealt with in that way. As a result, they tend to lead to mistaken conclusions rather more often than doing justice to the evaluation. As a rule, they display a prejudice which makes them turn a blind eye to indirect effects, or parameters such as usefulness, that cannot be expressed in monetary terms, or only arise in fact, from the interrelation of one factor with another. So it is then, that new technologies are assessed unfavourably in comparison with conventional technologies.

In making assessments, it is important to include secondary effects and those which cannot be expressed in monetary terms, and to consider the necessary organizational, staff-related and market-related measures that will accompany the change, in order to do justice to the "system character" of most investments. This becomes all the more essential as commercial organizations are increasingly dependent upon investment, and their performance on the market is often determined by the total systems they employ and the costs resulting from structural change. It is also important to carry out this evaluation in the light of future changes in the company's position, and not merely against the background of current production. Here it is necessary to consider such issues as what it will be worth to the company to have greater flexibility, to meet deadlines more promptly, or even to make better use of workshop qualifications. In the end, we must account for the fact that investments which appear to be cost effective in the short term can turn out to be thoroughly uneconomical in the long term (see pp. 119).

2.3.5   The Increased Demands of the World Market

The resolution of this situation in the present-day factory is a matter of extreme urgency, since the far-reaching changes in world markets which seem set to continue, have placed new demands on the manufacturing base. Apart from conjunctional variations in the trade cycle, there are frequent occurrences of saturation in the consumer goods market and consequently in the capital goods market as well, which together indicate a long-term trend towards stagnation. As the example of machine-tool manufacture clearly shows, the capital goods industry in particular will tend to stagnate because the capacity of its products increases faster than the range of their functions expands in value.

This evaluation cannot be altered by the existence in developing countries, of huge underprivileged areas with basic needs still unsatisfied. They have no purchasing power and therefore their needs do not constitute a demand in market terms. They are outside all market economies, and in addition, are tied so much to creditors that even in the long term they will not be in a position to develop their own economies to give them purchasing power on the world market. If they could succeed in doing this to a small extent, it could only be by disassociating themselves, if at all possible, from the events in the world market. They would need to form closed economies, such as those in China, Brazil and India for example; at least as far as important industrial goods are concerned. In consequence, there is unlikely to be a demand from the periphery.

The situation is quite different in the so-called developing countries such as Korea, Taiwan, Hong Kong and Singapore, to which industrialization has been exported. Even when they behave in a protectionist manner in the build-up of their industries, they have to rely on supplies from the developed countries. This demand is however, increasingly being offset by their own competitive exports. Consequently, it is the developed industrial nations that dominate the world market and will continue to do so well into the future. Of necessity, forecasts of low growth lead to fierce competition, which is aimed at squeezing out any opposition and building up the strength to dictate conditions to competitors. Success will come only to those who can meet the customer's requirements quickly, can guarantee deadlines and deliver promptly; and in addition can offer high quality and versatility.

The decisive features which will beat competitors, are therefore innovation, flexibility and the guarantee of delivery dates. Applied to the engineering industry, this means creating a broad range of high-quality products in an expanding section of the market, with short lead times and guaranteed deadlines. Future structures in production must address these demands.

Chapter 3

The Technocentric Route: Fossilized Taylorism

Despite restructuring procedures, which were so full of promise, short batch, customer-led production has run into difficulties. There are two obvious starting points from which this can be countered through the use of modern developments in computers. The first one is recognizing that an attempt to operate the highly capital-intensive equipment on a completely automated basis, is worth risking, at least for one period of the day. Doing this would increase the time for which it is used without additional expenditure on personnel, and thus reduce the hourly labour costs and the unit costs of production; a situation which is termed "the ghost shift" in business jargon.

The second one is based on the many different possible uses for computers in technical staff areas, where a great variety of activities are carried out, involving design, preparation, co-ordination, testing and so on. This variety becomes necessary when production is based on the division of labour, but it also proves to be inflationary, resulting in higher costs.

Management expects to gain two advantages from the use of computers in the technical staff areas. One is a reduction in the cost of company information processing, and the other is a much greater transparency in the production process, independent of the workshop personnel themselves.

As a consequence of this type of reorganization, there has for some time, been a proliferation of attempts to computerize, at least partially, certain product-related and order-related functions of company information processing by the use of data-processing equipment. However, this soon creates two new difficulties that are becoming more and more serious. Firstly, the shortsighted ideas behind technological and organizational re-design have created islands of fragmentation, when what is needed is a better combination of areas that were broken up by the division of labour, but which belong together from a functional viewpoint.

Secondly, the planning, implementation and flexible use of these systems, creates a new need for experts who are qualified in these fields, and can take all the related matters into account. This of course entails more costs and further dependence on such experts.

The technocentric concept of Computer Integrated Manufacturing (CIM), lays claim to the solution of these fundamental problems. Its protagonists predict amongst other things, that the factory will finally be rid of the wilfulness of human labour by using "artificial intelligence". We shall be pursuing the thread of this argument, with its possibilities and limitations in more detail in the following sections.

3.1 "The Ghost Shift": Forerunner of the Workerless Factory?

Let us first consider the risks involved in setting up "the ghost shift". The concept here is to view the introduction of the new technology as the key issue (Genschow 1983; Hammer 1983a: Spur and Ganiyusufoglu 1983; Warnecke and Steinhilper 1983a; AWK 1984d, e), whilst measures dealing with the organiz- ation of labour are limited to merely getting the residual work done. This is in line with the views of production planners who are shackled with Taylorism, and who see human labour as a source of trouble rather than a force of production, and therefore rely in the main, on the safe functioning of technological devices. According to this viewpoint, the following functions involved in production, should be automated. A prerequisite for this is that the machine tools involved are numerically controlled and have sufficient storage capacity for the programs.

The first activity to be automated is that of placing the workpiece in the working position and removing it again, following a sequence that can be altered at will. Clamping devices are used for this having interchangeable fixtures for prismatic parts and magazines for turning parts. When prismatic parts with relatively short processing times are being worked, the rapidly increasing cost of clamping devices limits the period over which it is worth operating under full automation (Warnecke and Steinhilper 1983a).

The second activity to be automated is the provision of the correct type and number of tools necessary for processing the workpiece. The complexity and variety of the workpieces mean that conventional tool magazines soon become inadequate, so if this is not to restrict the potential operational life of the machine tool, devices must be provided which permit the optional exchange of tools and subsidiary tool magazines (ibid.).

The third and final activity involves the provision of comprehensive equip- ment for automatic fault diagnosis, the monitoring of production, and recording operational data. This equipment is required to halt production immediately an unscheduled event occurs. The following are its major functions:

Ensuring a match between the program, the tools and the workpiece by means of coding

Monitoring the operational life of the tools

Diagnosing the state of all important components and recording the operational data

Managing the production requirements

Only when all these functions are carried out automatically, can production equipment be operated without using human labour. This can occur for a limited period of time only, by virtue of the restrictions mentioned earlier.

In a functionally divided workshop, fully automated operation for short periods of time, remains restricted to individual machine tools. That is, for as long as factory-wide transportation between them, cannot be automated econ- omically (we will come back to this later in connection with group technology). This kind of set-up is found in machine tools capable of completely processing parts; for example in machining centres, turning cells and occasionally in the production of sheet-metal parts (Genschow 1983; Hammer 1983a; Spur and Ganiyusufoglu 1983; Warnecke and Steinhilper 1983a).

The economic advantage can only be determined in individual cases of course, but examples do show what potential for rationalization there is in the ghost shift. In the case of prismatic workpiece manufacture in machining centres, comparisons have been carried out between conventional CNC machines, and the machines that can be fully automated for short periods. In each case it is found that, if there is continuous working during breaks and in the unsupervised third shift, the hourly cost is reduced by 30% (with each unit in use for 4215 hours; Hammer 1983a). In the case of parts produced in turning cells, this reduction has been 10% (with 3600 hours use; Spur and Ganiyusufoglu 1983). The tasks that remain for human labour during normal working hours are: removing finished workpieces; clamping down the blanks for machining (or emptying and loading the magazines); keeping the tools in order and installing them; preparing the programs and correcting any faults that might have developed, so that another ghost shift can be run.

In theory, no special kind of worker is allocated any particular task, since there is no plan in existence for designing work in a way that is suitable for people. In larger factories this allocation is carried out according to the principles of the division of labour, to which we have already been introduced. This means that the programs are then dealt with in the work preparation section, the tools are prepared by tool setters, the faults are corrected by maintenance workers, while the so-called operators merely clamp down and release the workpieces. What must therefore remain a mystery, is how to explain the somewhat stereotyped assertions made about it: assertions such as "This creates multi-faceted and responsible jobs requiring intensive training" and that the demands on the worker "are higher than those placed on the NC machine operator", and again that "severing the link with online operations creates a new freedom" (Genschow 1983, p. M7). Residual tasks have to be brought together to complete a job. Individual workers or groups of workers than get assigned to them.

Initial experience with production equipment, fully automated for short periods of time (and I refer here particular to turning cells and machining centres), indicates that the arrangement is technically feasible and economically beneficial, providing adequate organizational preparations have been made. It needs to be said though, that we are in no way dealing with the run-up to the totally worker-less factory here, as is evident from the narrow limitations explained above. Before that could happen, two further impediments would have to be removed.

Firstly, in order to ensure precision, the key dimensions of the workpiece from a quality control point of view, would have to be automatically measured both during and after the machining process, so that any necessary corrections could be fed back to the processing machine. Specific approaches to this problem are being worked on intensively, and some are already in operation. However, automated measuring devices that are flexible, and capable of general application in an integrated process, as well as economically sound, are still a long way off. Secondly, organizing the production process as required for the performance of routine tasks,causes numerous difficulties when applied to flexible automation and in transportation between the machine tools. To overcome these, three main development tasks would need to be undertaken:

Mechanical interfaces would have to be standardized so that the workpiece can be clamped with precision.

Convertible magazines would have to be designed for transportation, so that the workpieces are presented in the correct order.

A standardized coding system would be needed to identify the workpieces.

3.2   White-Collar Automation

3.2.1   Industrial Staff Work

The expansion in the volume of office work in the factory, and its separation into differentiated tasks, is caused by a number of factors. Considering first of all, conditions external to the company, extensive changes in the world market make a big impact. As long as there is growth in the market, a large part of the expansion can still be attributed to this fact, and to the additional administrative work it generates.

More recently however, there have been consequences arising from the efforts required in the struggle to squeeze out competitors, and which intensifies as stagnation sets in. It means that higher quality products, with a greater variety of customer requirements, have to be on offer (customizing). Moreover, shorter delivery times and greater adherence to deadlines are of increasing importance in competition.

Taken together, these new market demands bring about a two-fold increase in the complexity of company information processing. Firstly, they bring about increased complexity in design, development, quality control and the attendant planning of the whole procedure. Secondly, there is increased complexity in dealing with demands on the shop due to the greater variety of parts, the smaller batches and the increasing pressure to shorten lead times; notwithstanding the tendency to use the machine to capacity, so that stocks can be reduced and delivery dates honoured.

Taking mechanical engineering as an example, which is the largest branch of industry producing capital goods and the one with the greatest variety, these assertions can be empirically verified by considering the following factors:

1. On average, only every eighth quotation results in an order (AWK 1984a).

2. Out of the total throughput time of an order, 55% on average is attributed to pre-production work, such as material procurement, design and planning; 22% is attributed to the production of parts and 23% to assembly. As a rule, design alone accounts for 30% of it (Poths 1983; AWK 1984a). In addition, the number of cases where firms spend more on information processing than they do on processing the materials themselves, is increasing.

3. The growth of employment in mechanical engineering has shown a slight decrease in total employment in Germany since 1972 (there were 919 170 employees in the industry in 1960, 1 199 567 in 1970 and 1 090 278 in 1980), but there has been a considerable and continual increase in the percentage of white-collar employees compared with that of blue-collar employees. In 1960 the figures were 22.1% and 68.2% respectively, in 1970 the corresponding figures were

28.7% and 63.9% and in 1980 36.5% and 55.8%. The proportion of employees involved in technical office work has clearly gone up more rapidly than the proportion involved in commerce and administration, while the decline in skilled workers as a proportion of the whole has been less than it has for the remaining manual workers (Benz-Overhage et al. 1982, pp. 238f; VDMA 1982, p. 30).

There are some indications that the position of German mechanical engineering in world markets is no longer as secure as it was in the past. It does on average, offer the greatest variety of parts and the smallest batch sizes. Nevertheless, only about 3% of mechanical engineering firms have computer-based systems at their disposal for processing their orders. Moreover, only about 11% of the firms use CAD systems, whereas this figure is 18% in the USA and 25% in Japan (Baumgartner and Stöckert 1984).

Taken together, these circumstances are a challenge to firms producing capital goods to work out a tenable restructuring course between the Scylla of market-economy demands for quality and readiness to deliver on time, and the Charybdis of production-economy demands for shorter lead times and reduced costs. At the same time, management is faced with a problem in the field of white-collar work. It needs to avoid losing its independence from the experience, abilities and desires of qualified workers, which it gained when the planning and implementation procedures were separated, and extend that independence further if possible. We should bear in mind though, that this has its limitations, as we have seen.

The principle of technocentric production involves the same horizontal and vertical division of labour, the same formalization of work sequences and databases, and the same use of equipment that had already been applied in the workshop. This time however, the planners' work is also reorganized, but it will be evident in this case that a complete analytical examination of all the white collar work processes involved will itself come up against stubborn obstacles, and that the discrepancy between the analytically designed model and the actual work process will limit knowledge objectivization and thus require concessions to human labour. What could be more obvious therefore, than to start experimenting with partial processes first?

Fig. 6 represents the essential functions handled by a company's information processing system, and indicates the flow of data and materials between them. To make things easier to understand, the various methods for contending with the growing complexity of intellectual work will be considered. They will be dealt with in respect of two separate areas of work to ensure clarity, that is, the area of product-related requirements in design and production planning, and that of order-related functions in sales and production control.

3.2.2 CAD: Models of Models

An obvious approach would be to firstly direct attention to studying the organization of production, and to following the sequence of events in the design and development departments for adapting procedures and coping with capacity. On the basis of the analysis carried out, it should be possible to plan capacities and deadlines in the individual departments, and formalize the sequence of events. Measures of this kind have already been taken in a few instances. They make the

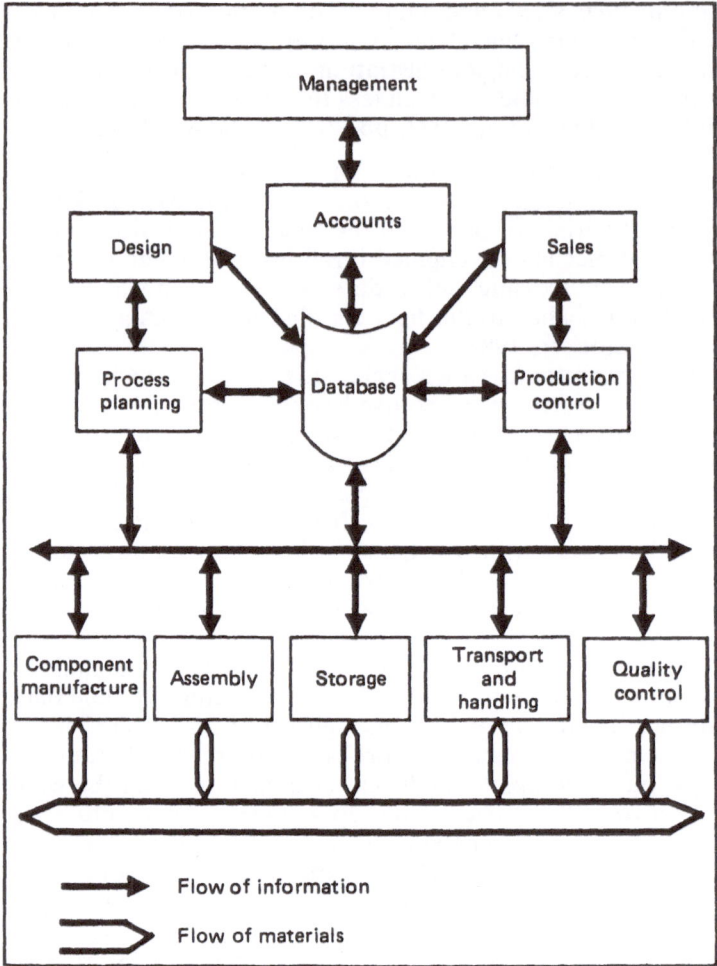

Fig. 6. Functional structure of the production process.

whole process clearer, more reliable and easier to control as regards time and cost (Brankamp 1981; Beitz 1983; Wiendahl 1983). However, having an overwhelmingly technocentric perspective, management looks upon increasingly cheaper computing power as the predominant element in the restructuring of design. The reasons for using CAD systems are the same old reasons as before, based on the imperatives of exploitation and control (Bechmann et al. 1978; Hesser and Rybak 1979; Beitz 1983; IG Metall 1983; Wingert et al. 1984; AWK 1984a).

Looking at the situation in the market where even shorter delivery times are being demanded, it is becoming more pressing for companies to reduce the design time of a customer order. Because of its large share of the total throughput time, design is seen as having a high potential for reorganization.

Looking at the production process itself, two ways for reducing costs need to be developed in tandem. Firstly, management expects to save on staff costs at the design stage, not directly, but by raising output with the same number of personnel. Secondly, although design accounts for only about 8% of the total costs (the figure is 7% in production planning) about 70% of these costs are determined during design (a further 15% in production planning). These facts provide an additional starting-point for lowering costs by searching more effectively than before for cost-effective design solutions (Poths 1983). In addition, it is expected that the use of computers in design will enable the quality of increasingly complex products to be improved through more precise calculation of the strength of materials and components.

Management is setting great store on the re-use of existing design solutions via the CAD system, so securing independent know-how which is separate from the design staff.

The use of the computer, which like every other machine is "implemented theory" (Bednarz et al. 1984), also presupposes that an algorithm of the work process can be produced. Such a model is only obtainable from precise analysis of the procedural sequences and knowledge of the materials. The more advanced the division of labour, and the more formalized the organization of the sequences of events, the easier it is to create such models. Effective use of computers is therefore to be expected first of all in those part processes which can be easily described algorithmically and which at the same time, promise particularly impressive efficiency increases.

The design process as a whole takes many forms and embraces a wide range of activities. Fig. 7 shows how design time is distributed. In the 1970s, there was a proliferation of contributions from computer science on the possibilities of using computing methods based on a general methodology for mechanical engineering. We are dealing with a discipline that has set itself the goal of "recognizing laws in the design activity and developing rules which are suitable for controlling this process rationally" (Hansen, quoted by Riehm 1982, p. 13).

The feasibility of consigning the design process to automation is evaluated very differently by some design scientists than by others. According to the most extreme view, the whole design process can be carried out automatically, right from the development of a functional specification to the design details. This is based on the assumption that all processes in technological systems can be reduced to a finite number of physical, chemical and biological effects, from which qualitative outlines and quantitative data can be produced.

Another view asserts that automation is not feasible in general for design problems, but holds that the whole design process can be supported with data processing by using interactive programs where the computer deals with the stages that can be handled in algorithmic form, whilst the designer concentrates on the creative stages.

Finally, there is the opinion that only certain partial design processes, ones that can be reduced to algorithmic description, are suitable for computerization, in particular the design of individual parts, the design of variants and calculation (Riehm 1982).

The euphorically high expectations of computers by computer scientists are soon followed in practice by a sobering reality. Computers were first in general use in the 1970s in the field of technical calculation, where their introduction was the most straightforward. However, according to Fig. 7, this accounts for only

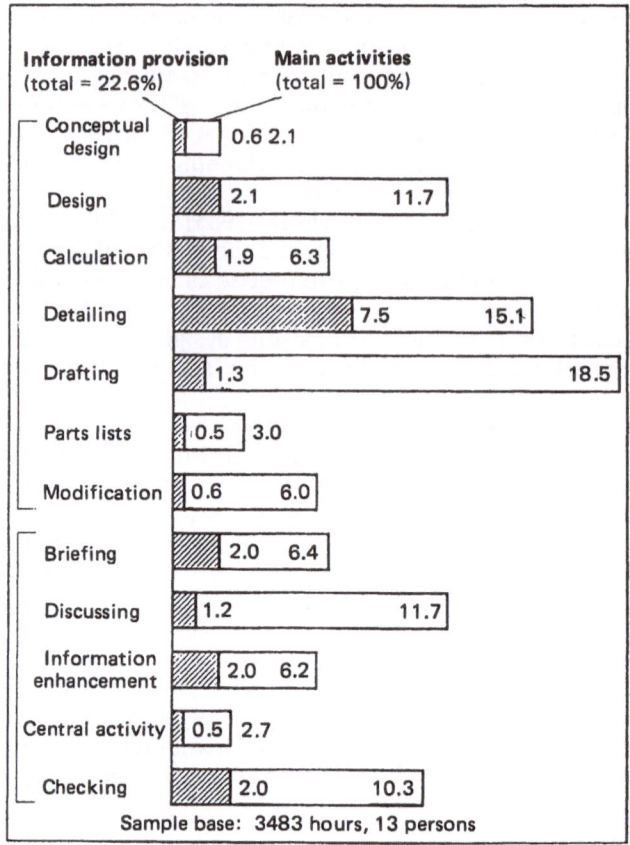

Fig. 7. Distribution of design activities. Source: Beitz, Hesser.

about 6% of design time, so the effect of rationalizing it is very small (the situation is similar in the case of parts lists). As a result, increased efforts were made to examine the processes involved in the drafting and detailing of engineering designs, in order to make them open to automation by computers. Since drafting takes up between 20% and 40% of the total design time, there is a great potential in these areas for rationalization. In some cases this time can be reduced to one fifth of the previous drafting and detailing time (Hesser and Rybak 1979; Riehm 1982; Poths 1983).

The use of computers began sporadically with CAD systems which were largely designed for specific tasks only. In the meantime, despite the great variety of applications, standard systems have become available. These are either in the form of program packages not specific to any computer, or as complete systems ready to use immediately. Summaries can be found in Grabowski (1982) and in Eigner and Maier (1983). Such systems are rapidly becoming widespread and worldwide growth rates of between 30% and 40% are expected every year. In the Federal Republic of Germany, about 400 CAD systems were in use in 1984, providing 1500 jobs, mostly in the design of individual parts and of variants

(Riehm 1982; Hatvany 1983; AWK 1984a). Attempts are being made to make up the gap by which Germany lags behind by creating a stimulus to demand in the form of a state subsidy in the order of DM 1000 million.

Work scheduling, which forms the link between design and production, has the task of using the drawings and parts lists prepared at the design stage, to work out the necessary instructions for the production processes. The following steps have to be taken for each production order. Firstly, there is the process planning, where the actual production processes and the order in which they occur are determined. Secondly, the resource planning, where the production equipment and means of operation are selected and thirdly, the time planning where the actual times are specified. The result is the work schedule.

In addition, NC programs may be drawn up for NC machines. From a more long term perspective, the necessary investments are also planned on a technological and organizational basis (Wiendahl 1983). Here too, efforts to rationalize are directed at shortening the planning time as well as improving and if possible, re-using the results of previous planning (AWK 1984b). It is already traditional for the major part of work-planning activities to exist already as a standard routine which contains standardized work sequences and formalized databases (for example, files on machines and operating procedures). These activities are predestined for automation using computers. The range of standard program packages offered in this field is correspondingly extensive. They are able to reproduce the entire methodology involved in planning a specific object and can generate work plans and NC programs on a fully automated basis. Some are designed for interactive use by specialists (Spur 1979; AWK 1984b).

3.2.3   Timetable by PPC

Dealing with orders calls for a variety of interlocking functions under the umbrella of production planning and control (PPC; or manufacturing resource planning, MRP). These are listed in Table 1 (Speith et al. 1981).

The job of programme planning is to establish which products have to be manufactured in a fixed period of time according to type, quantity and deadline, taking capacity and procurement into account. The way this is carried out depends on the type of production process involved. In the case of programmes where the products and their variants are well defined, such as lists of parts and work schedules, and where there is a small range of parts and longer production runs, we are dealing with the comparatively well structured problem of devising a sales programme based on sales expectations.

The situation is completely different in the case of capital goods production, which is short batch and customer led. In this instance, orders are placed solely on the basis of a quotation which contains the technical specifications, price and delivery date. These are fixed before the product is designed and before the drawing up of any parts lists and work schedules. Quotations can thus only be made on the basis of past experience in dealing with the problems of procurement, cost and capacity. This is a highly uncertain and loosely structured task (Scheel 1980; Wiendahl 1983).

In quantity planning (materials management), material requirements are established from parts lists. Existing stocks are used and procurements are made.

Table 1. The functions of production planning and control (from Speith et al. 1981)

Production planning	Production programme planning	Customer contract administration
		Demand forecasting
		Outline planning
	Quantity planning	Demand specification
		Stock control
		Procurement accounting
		Ordering
		Order control
	Deadline and capacity planning	Fixing deadlines
		Capacity requirement accounting
		Capacity synchronization
		Sequence planning
Production control	Order initiation	Release
		Document preparation
		Material provision
		Allocation of work
	Order control	Progress control
		Quantity and deadline monitoring
		Group capacity monitoring
Base data administration		Storage
		Modification

In deadline and capacity planning (time management), the capacity required is determined using the work schedule, and then matched to the capacity available. Finally, it is the job of production control to initiate the production process and get the batches through during a period that meets the deadline; simultaneously monitoring the progress in terms of quantities and timescales.

There are ever increasing demands on quality and performance which is reflected in the sheer complexity of the products, so the manufacture of these products is becoming ever more complicated, making it more difficult to deal with orders (Benz-Overhage et al. 1982; Hirsch-Kreinsen 1984; Manske and Wobbe-Ohlenburg 1985). The following example of a firm producing customer-specific individual goods, may give an idea of the extensive quantitative framework underlying this planning, and of the complexities to be handled.

This particular firm has 2000 employees of which 650 are involved in production. Within a planning period of six months, 1000 orders were processed, of which 2900 sub-assemblies were begun or completed. Of these sub-assemblies, 11 750 individual parts were begun or completed. To produce the parts which the company manufactures itself, 16 000 working procedures in 187 teams had to be controlled (Wiendahl 1983, pp. 250f). It is not therefore surprising that conventional methods are inadequate for the complex and comprehensive planning and control tasks of this type. Companies involved in short batch, customer-led production will thus have to suffer difficulties in meeting delivery dates, having to cope with high stocks (which may not necessarily include all the materials needed), long and varied lead times and frequent rush orders.

Difficulties are also created by the fact that the process itself causes a dilemma in sequence planning. This is because the requirement of working capital for higher utilization of capacity is in contradiction with its demand for lower stocks. For a long time, the market engendered more concern for the utilization of capacity, but recent demand has fundamentally changed this somewhat over-

valued concern, and it has now given way to flexibility in respect of the customer's wishes and greater readiness to deliver. Greater importance is now put
on the demand for shorter lead times, lower stocks and increased clarity in the
whole operative procedure. Whoever is the quickest at this, will make the running; at an appropriate cost of course (Kettner and Bechte 1981; AWK 1984c;
Manske and Wobbe-Ohlenburg 1985). It is particularly important in the case of
customer-led planning and production that at every stage the procedures are
related to the customer order in question.

There is indeed, a glaring disparity between the obvious need for effective
ways of planning and controlling customer-led production and the supply of
suitable data-processing systems. While it is true that standard PPC program
packages have been on offer by equipment manufacturers and software houses
for some time now (see Seliger 1978 and Speith et al. 1981), these packages do
not as a rule fit the bill. They are in the main, made up of planning sequences
worked out for program producers (Gerlach and Vortherms 1977; Scheel 1980).

The most important reasons for their lack of suitability are that they presuppose that complete parts lists and work schedules are available, and that
consequently, planning does not have to begin until a very late stage in dealing
with the order. Thus, computer support, which is so significant at the initial
stages, is left out or underdeveloped when the rough planning is being done, and
also when the quotation is being prepared. Moreover, despite being described as
modular programs, their structure is anything but modular and can be adapted
only with great difficulty and expense, to procedures organized in different ways.

Frequently, these programs were originally designed for batch operation, and
only later on were they provided with an interactive capability. Their programmed
decision algorithms (for example, automatic capacity planning), necessitate
compulsory sequences that cannot be reconciled with the need for flexibility in
reacting to boundary conditions that may arise. All this explains why only few
PPC systems are used in the production of capital goods. Even then, unless they
are tailored specifically to the company's requirements, they are mostly limited
to the short-term planning functions of production control (limited further to the
islands of part production) or to the loosely related management of materials
(Hirsch-Kreinsen 1984; Manske and Wobbe-Ohlenburg 1985).

The effect on work organization, of making use of computers in the manner
outlined above, means that the improvements due to restructuring as a whole
are kept within narrow margins, despite localized successes. The development
and use of current data-processing systems are geared entirely to the division of
labour. Their designers are now attempting to automate certain sections of this
division as completely as possible. However, the problems that arise in actually
getting this done often mean that work organization has to be adapted so that
human labour can be used to perform the residuary functions that cannot be
automated.

A comprehensive technological and organizational concept that would deal
with the interaction between human being and computer whilst having regard
for their respective strengths and weaknesses, does not yet exist. This holds
unless the present efforts by means of planning and certain forms of computerization, to reduce those features of the human being that have always been the
subject of complaint and reckoned to be the source of problems, such as unreliability, fallibility and inefficiency, are taken as such. It is now worth
considering these contradictions in more detail.

3.3 The Failure of Isolated Systems

3.3.1 Can Self-Will Be Expropriated?

The use of computers to rationalize intellectual work has become exceedingly complex. It presupposes, as in the mechanization of manual labour, as precise a knowledge as possible of the items being processed and of the procedural sequences in use; in short, the existence of a model of the work process obtained through analysis. Certain parts of this model are to be found in the formalized plans for structure and sequence and in the databases and data flows of the work process, which of course is already organized on the basis of the division of labour. These parts however, still have to undergo refined analysis so that they can be described in the form of algorithms and databases. Incidentally, there is logical consistency in the fact that this development starts out with so-called effective analysis and its generalization.

A new branch of intellectual work surfaces for this purpose. It involves system development, plans of action and the follow-up to these plans. According to a tried and tested model, this subjects the intellectual work of the employees involved to exactly the same methods of objectivizing their knowledge as was applied to the knowledge of skilled workers on the shop floor. Babbage's principle, and the principles of Scientific Management in the factory serve to transform the planner into the planned, the co-ordinator into the co-ordinated and the detail designer into a machine operator. Before they know it, a large part of their knowledge and ability, which they considered so indispensable, confronts them as something embodied in the programs and databases of the computer. Apart from just knowing about it in a formal sense, these people now actually experience the incorporation of their own human work into fixed capital.

This course of events, the historical prototype of which is in the work of the programmer, could be described as transitory polarization, and is becoming more and more common. At the practical level, with the original separation of planning and implementation, and the partial replacement of human labour with machines, there remained a large number of relatively simple residual tasks requiring few qualifications whilst at the planning level, a small number of relatively highly qualified activities arose. Lowering the value of the workforce in this way is, in accordance with the Babbage principle, the main economic purpose of the division of labour.

This situation will not last however. For every time a specialized qualification profile appears, and expands due to further horizontal and vertical division of labour, the high qualification demand leads to high costs and yet more division. In other words, highly qualified activities now experience rationalization pressures resulting in their own further specialization, thus becoming more easily susceptible to algorithmic description.

In the technical office, this phenomenon takes on the form of computer based systems for what are called islands of rationalization. There is production control, work scheduling and NC programming, calculation, provision of drawings and design variants. Sections of the former intellectual work are automated here, and at the lower end of the new qualification profile there arises specialized residual work in data preparation and input, and in the specification and arrangement of details on the screen. At the upper end of the new profile, the job of planning the whole system, introducing it and following it through demands high qualifications and very few people. At the same time the transparency and control of the work process is increased within the rationalization islands.

In the course of all these rationalization activities, the imperative of becoming independent from the knowledge and ability of skilled shop floor workers and of the specialists in the technical office, hardens into a sacred principle. Even if this were a measure that could be explained in completely rational terms, as was the case in the earlier stages of capitalist production when management had to gain the upper hand, and control the work process, it still does not explain this further determination to rationalize.

It is well understood that human labour is not only capable of good work that creates surplus value, but is also capable of exhibiting the antithesis of this capitalist requirement in the form of obstinacy. This cannot fully explain however, why it is that the principle has now taken on all the traits of a dogma. Even in engineering training, in written material and at conferences, it is continually reproduced and reinforced without ever considering the actual social background that once gave it credence.

Even in the minds of production engineers themselves, a pattern has been formulated which no longer permits the awareness of any alternatives. At first it must have seemed possible for them to assert themselves using the technology they had designed, and strengthen their own position of power in relation to manual workers, as Taylor's own experience seemed to confirm. Now however, this same technology must appear to them to be the cause of their own inferiority. For the more deeply they are convinced that efficient and error-free performance can only be achieved with advanced planning which is isolated from the activity itself, the less they are able to accept the practical deviations from the plan as anything but human imperfections in badly planned actions. For them, the imperfection itself is their worst offence. They are prevented by their misguided conviction from putting up a fight against the supposed imperfection of the computer based plan and they actually see themselves as the defect in it. Consequently, their persistent view of labour as design defect, and in particular as a source of problems to be eliminated if at all possible, results not from their superiority over blue-collar workers, but from the inferiority they feel in the face of the machine system. For them, the prime imperative is the compulsion to think in machine terms. They are in the peculiar state of mind which Gunter Anders diagnosed a long time ago as "Promethian shame":

It wasn't because he could no longer bear anything that he hadn't made himself or that he wanted to make, but because he himself didn't want to be something that was not made. It wasn't because he was indignant at being made by others (God, the gods or nature), but because he was not made at all, and as such was inferior to all that which he made. (Anders 1973, p. 25)

It is evident that with this thinking, there is little room for organizing the manufacturing process in a way that is suited to people as well as to production. The whole rationalization process involves objectivizing as much of the work as is possible within economic limits. As much empirical knowledge as possible is described in the form of work-sequence algorithms and databases. These are then transformed into planning procedures. Thinking in machine terms means that things cannot be allowed to stand still until as many areas of production as possible have been automated.

But this technological ideology is itself only the reflection of something else: the reality of the capitalist production method. The mistrust of engineers towards human beings is a manifestation of the mistrust of capital towards human labour. The elimination of human inadequacy and of insecurity is the engineers' expression of the attempts of capital to minimize its dependence on human labour by strengthening its control over production. (Noble 1979, p. 18)

There are several fundamental reasons why these attempts can only solve to a limited extent the basic problems of flexibility in reducing stocks and lead times.

Firstly, job shop production, organized on the principle of functionally divided jobs, remains untouched by these rationalization measures. However, this principle is the main cause of long and sharply varying lead times and thus of high stocks as well. The use of computer based systems for production planning and control can reduce but not eliminate these problems.

Secondly, the principle of separating planning and implementation is based on the need to obtain an abstract model of the process. The disparity between the model and the process itself is in turn, a reason for the actual work sequences differing from what had been planned. This discrepancy is made worse by the planners themselves, who reject experience when they carry out their work. Consequently, these planners develop their internal representations from their own model planning and building, and these guide their perceptions and actions. As a result they incorrectly perceive and interpret any difference between plan and reality (Troy and Schüpbach 1984).

Thirdly, the islands of rationalization created, are not particularly compatible with the information processing areas in the same company. Although split up according to the division of labour, they are of course functionally related. This incompatibility necessitates additional co-ordination; in particular, the repeated preparation and input of data to cope with the transfer between isolated data processing systems. The whole procedure is both error laden and very costly, absorbing the earlier gains made within the islands through rationalization. This incompatibility makes the integration of the different systems a matter of urgency. It does however meet with design difficulties, because the algorithms and data structures associated with each system have been independently formulated.

Spurred on by the shortsighted and short-term interests of capital and its urge to secure control, the first great wave of rationalization in intellectual work advanced with tremendous force. It is now smashing against the wall of yet more difficulties. They culminate in that irony of automation which originated with production engineering, that is that the systems developer, in attempting to eliminate the "unreliable and inefficient" human being, substitutes his own errors as the main cause of operational problems, even though he still needs this human being to perform the tasks he can't automate (Bainbridge 1982, p. 151).

3.3.2   Coping with Uncertainty

The difficulties referred to above can be substantiated empirically. In order processing we can start with computer-based production control systems. As a rule, their introduction brings about no real radical effects. There are however, some conflicting but clearly recognizable changes in the organization of work, affecting the allocation of authority, the requirement for qualifications, the work-load and the extent of control (Benz-Overhage et al. 1982; Hirsch-Kreinsen 1984; Manske and Wobbe-Ohlenburg 1985).

In conventional production control, the release of orders and the specification of deadlines and their monitoring are assigned to the control personnel, whereas the arrangement of the sequences and the distribution of labour are determined in the workshop. In the customary "foreman economy" the foreman's central task of allocating the work according to qualifications, technology and deadlines, left the workers with considerable freedom to arrange the way they carried it out. This, together with the piece-rate incentive, encouraged speed, and thus brought about important achievements in workshop design with regard to the preparation of machines and the positioning of tools and equipment.

It is the avowed intention of management to eliminate this chaos by using data processing, rather than let it be handled through improvization. This intention has been adopted by production control as part of an overall computer-based plan to allocate labour, co-ordinate deadlines and capacities, determine the sequences and monitor everything via feedback systems or data-capture terminals. In this way, the production control department experiences a clear expansion of its authority and an increase in the potential for control, whilst the number of comparatively poorly qualified workers pursuing deadlines goes down sharply. The work of the remaining specialists no longer includes the former routine activities; rather it concentrates on the main control functions, being highly formalized and adapted to data processing conditions. The qualifications required for this work vary, according to which sections of the planning system have been computerized.

In contrast with the original division of labour in the workshop, the foremen are no longer involved in a great number of control and co-ordination tasks and thus lose their previously dominant function. At the same time, the workers themselves lose the considerable scope for organization which they had earlier, mainly because computer-based work planning involves more precise specifications. In consequence of this, an important area of capability is no longer needed even though machining work is increased.

The effects which management intended in all this have certainly come about to a certain extent. By such means there is a clearer insight into the activities within the factory, making it easier to control the work sequence and performance levels and keep to the work schedule. For example, there have been some reports of workshop stocks being reduced by 20% as a result of the change.

There is another side to the coin, for attempts to make radical use of the mechanisms for control, have led to equally stubborn resistance, bureaucratic behaviour and all sorts of ploys to get round it. In addition, the constraints on planning of short batch, customer-led production are obvious. Design deficiencies (and subsequent changes), machine failures, shortages of tools and equipment, slippages and timescales that can't be made precise together with fluctuations in performance, all lead to the most detailed plan being flawed by events after only a few hours. Costly revision or even renewed efforts at improvization have then to be made. Neither is this changed fundamentally by using real-time operational data, so in a number of firms it is company practice to control production in a way which is not completely tied up with overall planning (Fröhner and Duda 1979; Benz-Overhage et al. 1982; Hirsch-Kreinsen 1984; Manske and Wobbe-Ohlenburg 1985).

The greatest variation is exhibited in batch control. A particular supply of production orders with specified deadlines is sent to the workshop as a result of computer-based planning. The workshop, supported if need be, by its own computers with their custom built procedures (Spur et al. 1983), is left to exercise the division of labour itself and to determine the sequence of events. The capabilities of human beings to deal with faults skilfully and on the spot, are thus consciously made use of (Manske and Wobbe-Ohlenburg 1985).

Difficulties proliferate however, in the area of customer-led production, for provisional commitments have to be made which will become binding when the order is accepted (or even beforehand). These commitments, mainly concerning materials and capacity requirements, have to be made without the advantage of any detailed knowledge. Only after the order has gone through the design and work scheduling stages can this information be known more precisely, for as long as the computer-based management of time and materials is isolated from the rough, or initial planning, there is no other option left to production control but to attempt to smooth out the difficulties afterwards. As long as rough planning does not have reliable estimates available to it at the quotation stage, it can only cope by referring to the way in which former, similar orders were handled in the past. This means that there is more or less blind approval by production control when orders are accepted.

The actual interconnection between the functions involved in dealing with orders means that it doesn't cope very well with isolated rationalization. What is urgently necessary here is integration of the divided areas into one unified system of computer-based production planning and control. Given the characteristic features of order processing, that is that the planning data is refined in stages as this process evolves and that it is used again and again (and altered) in various functional areas, this integration cannot be achieved by linking together existing partial systems as an afterthought, but only by designing a fundamentally different system. An essential component of this new design is a common database, accessible interactively using programs specifically related to the function in hand.

The effective use of computer-based systems for work scheduling and NC programming presupposes standardization throughout. Although these systems are probably structured from the outset, it is still a job of considerable organizational complexity. On the other hand, this formalized and regimented method of working and the relatively large proportion of routine activities it involves (searching for technological data and calculations) make it possible to automate the greater part of the planning which is not customer-led, leaving comparatively few planning steps to be carried out in the dialogue between the specialist and the computer. Consequently, the technological and planning aspects of work are sharply reduced, and the need for work schedulers also diminishes. All that is required now to implement the work schedule, is a knowledge of systems operation.

The drawback of working with these work scheduling systems is just as clear. There are complaints about them being out of date because the time lag when planning is separated from production. Discrepancies between the planning specifications and the current production situation usually necessitate around 30% re-planning. In addition, there is no desire to change production conditions for clumsy adaptations of programs and data requiring constant maintenance.

Looking to the future, consideration is being given to the complete automation of routine planning when the work is not dependent on special orders, and the transfer of customer-led planning back into the workshop production arena. Traditional production planning is clearly in a bad state and new tasks are now

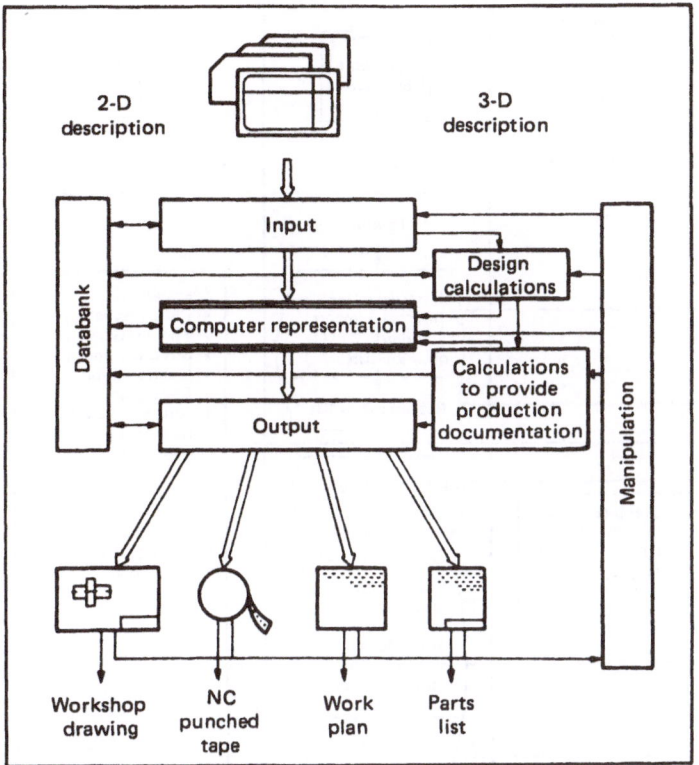

Fig. 8. Structure of a CAD system (from Wiendahl 1983).

being assigned to it, such as advice on product design so that design can be matched to production (Beitz 1983; AWK 1984b). This approach is somewhat dubious when we consider that work schedulers are not always familiar with the production process itself, but only with models that are of necessity incorrect.

3.3.3 Does the Designer Shape the Design or the Design Shape the Designer?

The introduction of CAD systems into design (Fig. 8) brings about a new division of labour, quite independent of the previous one where the design of ideas for products and sub-assemblies was a separate activity from the production of detailed drawings and specifications. This new division removes the tasks of systems planning and maintenance from the work done by the designers when they use the equipment. It is dealt with in this way because management recognizes the difficulties inherent in the previous arrangement such as confusing tenders and unsatisfactory system groupings. Unlike the use of computers in calculation, this method of dealing with CAD opens the way for further rationalization and is therefore important to company strategy (Schlagenhauf and Schaffitzel 1983; Schaffitzel and Sellmer 1984).

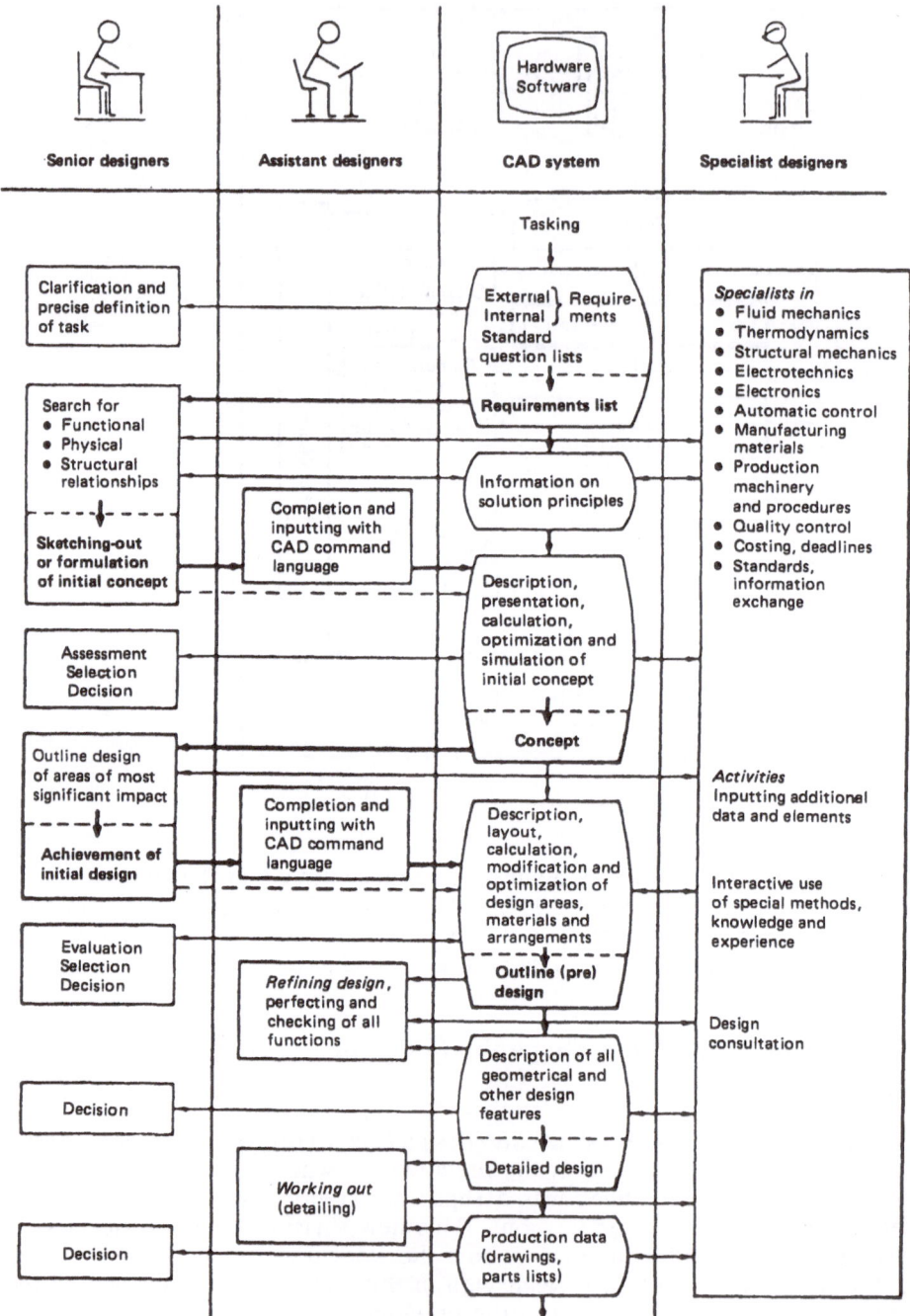

Fig. 9. Conceptual model of the design process (from Beitz 1983).

At the same time, this new division of labour lays the foundation stone for Taylorism in the design process. Designers and draughtsmen have to take part in planning and installing the system, in order to take proper account of the demands made on the CAD equipment. However, as the training arrangements and training principles show, the designers' knowledge of the system is restricted to whatever is necessary for dealing efficiently with it. Planning the design process itself, which was formerly part of the job done by designers and draughtsmen, now becomes the major task of those responsible for systems design. Answerable directly to management, they have to systematize the use of the CAD systems themselves, on the basis of a generalized design methodology. They are also responsible for the design and development of the CAD systems themselves, both in a technological and an organizational sense.

A structure designed on the principles of "Scientific Management" quickly arises on this foundation stone. Beginning with design methodology and the properties of CAD systems both present and future, the design process is now to be dealt with on a CAD system, with a senior designer leading the project, several assistant designers and some specialist designers (Fig. 9).

In this process, it is to be left to the co-ordinating senior designer to establish an initial concept and work out an initial design. The responsibilities of the assistant designers are to input these initial design decisions into the computer with the aid of CAD command languages and to carry out refinements in dialogue with the CAD system. The specialist designers are concerned with interpretation, variation and simulation of the concept and they deal with the initial computer "blueprint". The whole sequence of this division of labour is enforced by a sequence logic, contained within the CAD system itself. Before the start of a project, the senior designer has to assemble the relevant sequence logic by combining standard work procedures. (Beitz 1983, p. 65)

The essential feature of this organizational structure is its central information and database. The CAD workstations are combined so that "once design data and design structures have been input, they can be made use of by all those involved and can also be checked by the senior designer" (ibid. p. 66). Moreover, it is seen as particularly economical to input design structures into the CAD system, so that assistants who are specialized in particular areas can deal with the variations.

Although this way of organizing the design process is considered an ideal one, it cannot be put into practice at present in this particular form owing to deficiencies in the technical integration of CAD programs. Nevertheless, many of its characteristics and components can be found in the systems in use today, where there is a similar line of development. A few systems, such as those used in the design of heat exchangers and pistons, have managed to deal with the entire design process in this way and link it in with work scheduling. The technical limitations of the CAD systems currently on offer restrict their application to specialized areas that have already undergone a form of division of labour; a division which is either product related (circuit diagrams, pipe diagrams and gears) or process related (design variations and the production of drawings).

All this presupposes that precautions have been taken to formalize sequences and that transfer rules have been specified. These lead the designer and detail draughtsman concerned along their own paths of specialization, with far-reaching consequences for qualifications, workload and the content of design work in general.

The CAD system is a new tool for transforming work that previously involved complete forms and functions, into analytical work involving the combination of elements (either fundamental elements or parts). Previously, the drawing represented the whole design activity made up of its simultaneous and associated parts. Now, the completed design is made up using abstract commands that unify the geometrical elements. This is done by means of the linear and sequential procedure of the machine algorithm (Riehm 1982; Bednarz et al. 1984; Wingert et al. 1984).

The object of the work and the means of doing it, thus come into conflict, and this can only be resolved in the mind of the systems user. This demands additional efforts and is burdensome. For example, the use of CAD systems depends on the precise planning and preparation of the work sequences in order to guarantee efficiency. The numerous routine activities that were necessary before, now disappear and are performed by a computer program. They are however, replaced by new routine activities needed for dealing with the CAD system itself, such as the strictly formalized input procedures and procedures for checking and amending the results.

Knowledge gained through the experience of drawing up and assessing designs, which was in great demand at earlier stages, is going to waste. Instead, an ability in systems analysis and the capacity for abstraction in the form of symbols and commands is what is now required. The system determines the sequence and rhythm of the work as well as the method. The work thus intensifies, demanding increased concentration on the part of the operator. Since all the essential stages of the work, including erroneous inputs, are initiated at the CAD workstation, the system provides a method for control through the monitoring of performance levels. In addition, the way in which the design work is carried out on the CAD system means that social communication is diminished in favour of machine communication (IG Metall 1983; Wingert et al. 1984).

In contrast, the protagonists of CAD technology insist that the use of CAD systems is associated with liberation from routine tasks, and that it encourages the development of creativity. Apart from the fact that continual demands for creative performance could easily prove unwelcome, all the feedback obtained from industry gives the lie to this assertion which appears to be true at first sight. For one thing, creativity presupposes the possibility of encountering the unknown and the unexpected, and the only way the mind can be stimulated to explore these new paths is against a background of past experiences. It is common knowledge (substantiated by experiment) that people who have to follow strictly regulated sequences for long periods of time, begin to exhibit ways of thinking and behaving which have ossified through routine. They lose the ability to deal with the unexpected, and to develop new ideas for solving design problems.

CAD activities have to follow strict rules and deal with abstract symbols and commands corresponding to the sequential plan embodied in the system. This plan attaches great importance to "user control" and can cause creativity to wither rather than develop. Indeed this is admitted by design specialists when they recommend that:

The senior designer should be separated from the operational rhythm of the CAD system since it is hard to reconcile creativity with being too closely linked to an automated system. As before, designers are needed who are in a position to create original solutions by conventional means, on the basis of broad specialist knowledge. This ability must be preserved in the interests of retaining an innovative force within the company, if it is not simply to operate within the range of solutions and methods already stored in the CAD system. (Beitz 1983, p. 66)

So, the sweeping statement about freedom from routine and the development of creativity, has to be understood as just a slogan for promoting aquiescence, rather than an accurate description of the actual state of affairs.

Design solutions (which were previously in the form of drawings and parts lists) are used in many areas of the factory and are supplemented not only by work scheduling and NC programming, but also by time and materials management. Quite often, former results are called up and just modified for immediate use. This is a common occurrence in mechanical engineering. This re-using of design information is the incentive to make it all part of the system. The results can be stored as data structures and transferred via the computer to the functional areas requiring them. In fact this seems to be the only way that the complex, error-prone inputting of the data can be avoided (Krause 1983; AWK 1984a).

There are two main prerequisites for doing this however. Firstly, the products must be structured in the form of subassemblies, and these must be standardized and classified: a task particularly suited to computers (Beitz 1983; Dietz 1983). Secondly, the various computer based processes have to be integrated in one computing system. This process proves to be extraordinarily unwieldy (Eigner 1983). There are already systems where it is possible to create data processing chains which connect various functional areas. They are mainly used for linking NC programming systems with CAD systems (Eigner 1983; Hellwig et al. 1983; Marktübersicht 1984; RKW 1984). However, they are of course unsuitable as a means of solving the general integration problem.

All in all therefore, it is clear that the moves toward reorganization in industrial intellectual work have generated separate islands of computer operation, and now it is a matter of urgency that they be linked together. This would be the final step in the integration of the whole production process organized according to the division of labour. However, it is hampered technologically by the isolated development of the islands which were not structured with integration in mind. Alongside this, the complex and unpredictable production process is stubbornly resistant to the use of computers. As at earlier stages, qualified human work is needed far more than expected. Technocratic rationalization has thus reached an impasse and to escape from it we must have a fundamentally different reorganization strategy (Sigismund 1982).

3.4   The Second "Heroic Phase": CIM and Expert Systems

3.4.1   CIM and Segmented Work

The solution that technocentric production provides for dealing with difficulties arising from the separate areas of specialization is called "computer integrated manufacturing" (CIM). Up to this point, it has been possible to base our account on existing systems and on measures already in operation, as well as on empirical findings from their actual use. From now on however, the discussion will be

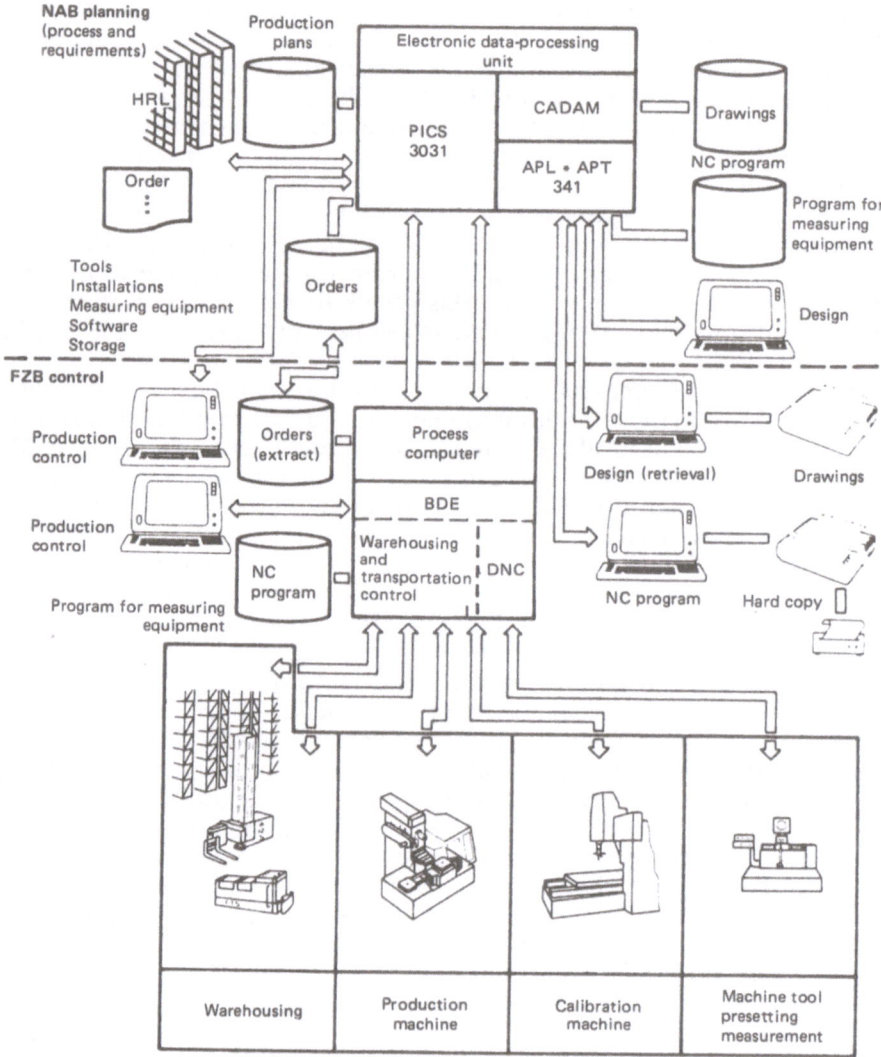

Fig. 10. Example of a partially implemented CIM system (courtesy of Messerschmidt-Bölkow-Blohm).

based on plans which are published but not necessarily implemented, and which have arisen from developments that have made some progress toward partial realization of the CIM concept (whose range and unequivocal claims leave nothing to be desired of course!).

The technical and organizational problems that get in the way of integration as a result of isolated specialisms in technocentric development, can be characterized as follows:

Programs and data structures are closely interrelated since the former determine the latter (and vice versa). Consequently, developments of independent data-processing systems lead to a dual incompatibility of data and programs. The technically feasible, albeit complex method of using translation programs to match up the different island programs and data systems, and then combine them (see for example, Müller et al. 1983 and Eversheim et al. 1984), leads directly to the next impasse. For this integration, when completed, then baulks at any attempts to link in further modules (notwithstanding problems of portability), especially at the attempt to exchange one integrated module for a better one which is not already part of the system. This integration procedure proves to be exceedingly resistant to innovation. Fig. 10 illustrates an example of this partial implementation of the CIM concept.

Further, the market situation in product-related information processing is itself blocking this way towards complete systems integration. From the supply angle, there are numerous software houses whose knowledge of technology and systems, makes them particularly well suited to the job of developing and introducing efficient CIM modules, that is, in addition to the large computer manufacturers who offer comprehensive packages. Moreover, many equipment manufacturers have recognized, that in order to take part in this area of development, they have to offer program modules for the operation of their installations, that can be integrated into the system. In order for this to work effectively, the partial systems must be compatible. From the user angle, we are dealing with an industry characterized by small and medium-sized enterprises, whose investment level and product diversity compel them to build up their system in stages, using CIM modules of several different origins (Sigismund 1982).

The market for automated systems has now assumed considerable proportions, and there is fierce competition, particularly between the international electrical companies. For General Electric, one of the leading suppliers of automation technology, the volume of trade in Western Europe in 1985 is estimated to be 2000 million dollars and in the USA 4000 million dollars. The predicted figure for 1995 in the US market alone, is 10000 million dollars (VDI-Nachrichten, 1985).

Bearing in mind these circumstances and requirements, an adequate CIM systems architecture which aims to combine all the separate systems in each functional area of a company into one integrated computer system (Fig. 11) must, as the bare minimum, possess all the following structural properties (CEC 1984).

Firstly, it must have a common database in which all the data structures associated with the system are stored and maintained. The functional area programs access this database, use it, and when necessary modify it. This ensures that each item of data appears only once, and that all the data is up-to-date, and consistent with the interlinked functions. Among other features, this database comprises:

Internal models of designs (drawn up at the design stage and used for example, in work scheduling)

Parts lists (drawn up during design and used for example, in production planning)

Work schedules (drawn up in work scheduling and used for example, in production control)

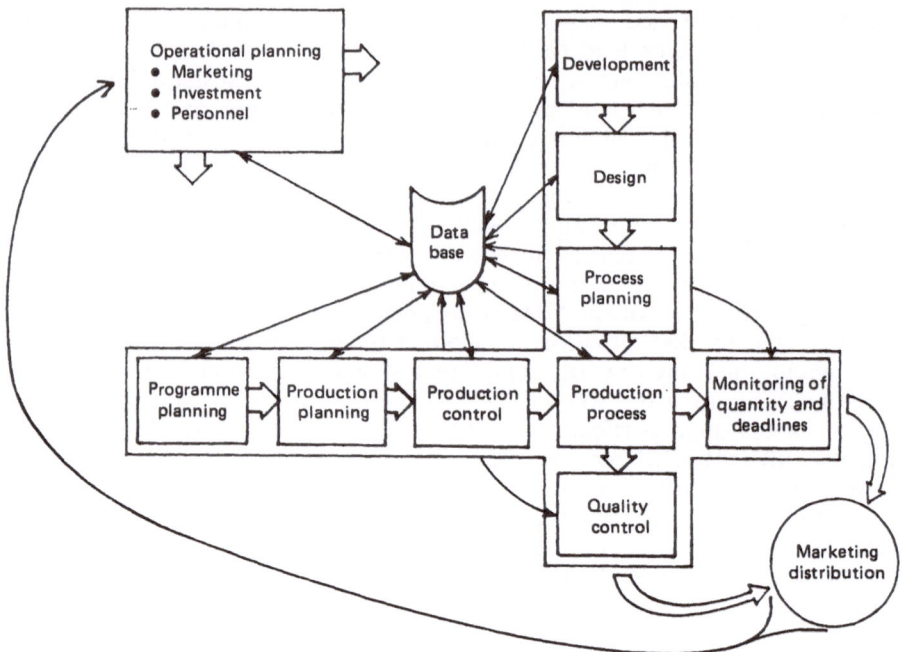

Fig. 11.   Horizontal and vertical integration in an information system for manufacturing.

Fixed data on parts and sub-assemblies (drawn up and dealt with in materials management and used for example, in design)

Current data on orders (which is related to the immediate situation)

Fixed data on customers and suppliers

Secondly, there have to be agreed cut-off points between the database and the programs in the functional areas. Only on the basis of such agreements can data be transferred between systems and modules of different origins.

Thirdly, a local network has to be provided whereby data is transferred between the separate partial systems according to an agreed protocol.

Only when these fundamental elements have been incorporated into the CIM architecture, does it become possible to have all the systems of various origins working together in the factory. Only then will it be possible to reap the full benefit of economic efficiency, transparency and control which are promised by technocentric production.

At the same time, this type of integration creates new opportunities for making further areas of intellectual work open to the use of computers, especially in the design office, which previously baulked at the idea of automation. Thus, untroubled by earlier failures, hopes are directed at bringing the use of computers closer to the stage where they can be used for original design, by using the stored geometrical models and the improved facilities for manipulation these systems offer.

The scale and complexity of computer-internal design models place demands on storage, administration and handling that exceed the capabilities of currently

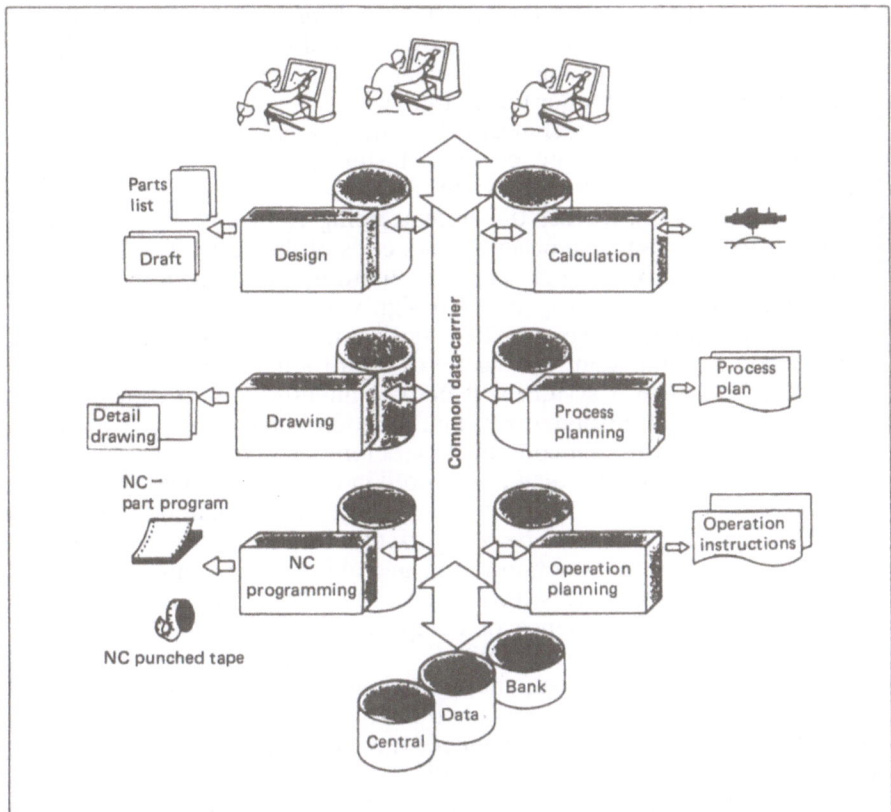

Fig. 12. Advanced production system (APS).

available systems. Therefore, great efforts are underway to develop efficient technical database systems and geometrical modellers that can easily prepare these design models for processing by function-specific programs, for example calculation, drawing, work scheduling and NC programming (Bjørke 1983; Krause 1983). Moreover, the large number of systems already available makes the formation of links within and between the individual functional areas an essential activity. For example, different CAD systems can be linked, but this is only an intermediate solution, born of necessity and permitting rationalization through integration, even without CIM design becoming a reality. An example of such a system is illustrated in Fig. 12.

The data structures associated with orders, are somewhat simpler. This fact, together with the urgent nature of production economics, has led to the development of ready-to-use PPC systems. These are a partial solution to CIM architecture, supporting all order-related functions using a common database in dialogue with the computer (see for example, Kölle et al. 1984).

Since the whole production process is placed on a new technological and organizational footing by computer integrated manufacturing, new ideas on the redesigning of structures and the sequence of events must be made integral to the CIM concept itself. We have a long way to go before these ideas are elaborated to the same extent as plans to advance the technology (Sigismund 1982). In particular, few fundamental changes can be expected in the deployment of human labour, given the thrust and inner logic of the whole development. Starting with the division of labour in the technical office, the imperatives of capital and its need for absolute control (better, cheaper, faster and yet more independent), drive this increasing polarization ever onwards.

In the course of this process, the human being is further driven aside to perform peripheral tasks which occur at both ends of the scale. The tasks at the lower end of the scale arise as a result of automation, and at the upper end they arise as a result of particular aspects of design work being open to the use of highly developed computer systems. The logical conclusion of this is an integrated computer system encompassing the whole of the production process.

The drawback of such a system will be a segmented labour structure with human labour performing isolated residual functions that will only combine to form a notional whole through computer modelling. The principles of scientific management, can be seen once again, fossilized in this arrangement. But despite management's high expectations of computer integrated manufacturing, here too there is a problem that has to be recognized before there can be any chance of success. To the extent that the knowledge and abilities of the human being are objectivized in computer systems, their scope and complexity increase dramatically. Despite this, all experience of complex and comprehensive computer systems teaches us that they are unreliable and not easily comprehended. The way they are designed and used, the dependence of their development and maintenance on the division of labour, and the heuristic methods on which they are founded, mean that there is no longer any single author who could fully explain the entire system (Weizenbaum 1977).

At present, more than two-thirds of the expenditure on programs is spent on their maintenance. All attempts to formalize program writing, fail to bring about any fundamental change in this situation whatever the quality of the software. The meaning of a formal expression depends entirely on the context in which it is used, but that context can only be described informally. Therefore, formalization has to rely on the human intellect in order for it to be put to use. Concrete problems cannot be converted into program specifications by formalizing. Neither can the correctness of the programs derived from them be verified (Keil-Slawik 1985).

Unable to rid itself of unreliable human beings through the mechanistic implementation of algorithms, management finds itself caught up in software production that relies on qualified specialist knowledge, and yet can guarantee neither correct nor reliable programs. It is ironic that in its urge to overcome the wilfulness of the human being, management has created for itself an impenetrable morass of unreliable and imperfect software and equipment, which is no easier to comprehend than before, and things are getting increasingly bogged down. More grotesque consequences of mechanistic thinking can scarcely be imagined.

The prospects of using fossilized Taylorism to overcome the existing fundamental problems of the flexibly automated factory, are as slight as were those for the technocentric division of labour at an earlier stage. The gulf between the planned course of production and what actually happens still persists, and we experience the ever decreasing ability to react swiftly and appropriately to unforeseen circumstances (capacity to cope with uncertainty). In light of these rather depressing factors, the claims made for the fifth generation of computers and for expert systems in general, would have us believe that they provide the basic solution at just the right time.

3.4.2   The Claims of "Artificial Intelligence"

With the fifth generation, of which nothing tangible exists yet, completely new paths will be explored. So far, there are only ambitious plans, a few modest laboratory experiments and some development budgets which amount to some billions of pounds worldwide (Moto-Oka 1982; Feigenbaum and McCorduck 1983). In contrast to conventional computers, the emphasis in the new generation will be on logical efficiency, not arithmetical efficiency; on having knowledge bases at one's disposal, not on rapid access to comprehensive databases or databanks; and on dealing efficiently with knowledge, not on rapid data processing. The new systems are no longer to be termed "data-processing systems" but "knowledge-information-processing systems" (KIPS) to correspond to the new development, and their performance is measured according to the number of logical conclusions they arrive at per unit of time, rather than the rate of numerical operations completed. In this connection, the term "knowledge base" means the totality of all facts relevant to a specialist area, together with the associated rules and assumptions.

The transition to the fifth generation is undoubtedly a radical qualitative change, a technological quantum leap should it succeed. Its intention is to go beyond quantitative changes which have predominated so far, such as the doubling of capacity and halving the costs of computer systems every two years, and instead, set off for pastures new. The intention of this fifth generation is to make feasible the efficient use of so-called expert systems and perfect the interaction between human being and computer by using natural language.

An expert system is made up essentially of three parts. There is the knowledge base, the inference engine and the user interface. The knowledge base contains the facts required for a particular field of work (the base having been designed with these facts in mind), as well as the assumptions and rules which apply to them, in much the same way as in text books and specialist publications. It includes an administration system, which is used to provide access to the elements of the knowledge base, to alter them and to add new elements. All this then, is what differentiates a knowledge base in an expert system from a database in computers of earlier generations. The database was just a store for data and data relationships arranged for sure and rapid access. The inference engine makes use of the simple procedures for drawing conclusions that are common in the specialist area concerned. These standard procedures can be used to combine various facts, deal with them, and assess them in the light of the data available. Finally, the user interface enables the user to put questions to the expert system in the most natural way possible, and to receive comprehensible answers to them.

The pilgrims on the road of technocentric development, are now hoping that these new technological breakthroughs and promises of "artificial intelligence" (AI), which are in any case, no more than promises, will once again ease the rigidity of the factory which has been fossilized by Taylorism, and do it without having to make concessions to the workforce. Not only that, they are even hoping to be able to eliminate human labour almost entirely.

In spite of insisting that increased flexibility and shorter production sequences have been achieved, they themselves have quite probably ceased to believe in the ability of conventional programming methods to avoid incurring yet more problems. Otherwise, one is at a loss to explain the absolute determination with which they attempt to instil "intelligence" into machines. "If we look upon intelligence in general, as the ability to adapt to new situations and behaviour, we can define machine intelligence as the ability of a machine to adapt the way it directs a process to the varying structural and other parameters" (Spur 1984, p. VI).

But there is a further reason for listening to the orchestration of those involved in AI research. In the course of flexible automation, the dynamic spiral of specialism continues to turn. As it turns, an increasing amount of expert human knowledge forms at the upper pole, driving up costs and creating new contingencies. It seems all the more understandable that management should want to incorporate this knowledge, reliably and cheaply into expert systems, in order to have it at their disposal without recourse to any human expert.

The coupling of various databases and information systems with appropriate expert systems would remove considerable deficiencies in the current production system. Machines which process knowledge and are capable of learning will be able to store their own operational experience and compare them with the experiences incorporated from their programs. Here we have an important possible innovation: the availability of the knowledge and experience of employees, even when they are absent from, or have left the company. (ibid. 1984, p. VI)

The intention is to introduce expert systems throughout the whole factory in order to make integrated manufacturing systems as efficient as possible. "Through their use, knowledge cells can be constructed for design, development, production planning and assembly as well as for quality control and sales" (Spur 1983, p. 12). It is also the intention that machine intelligence should now tackle single-handed, all the faults and "accidental" deviations from the plan, such as inconsistent properties of materials, undesirable changes in the surface and form of the workpiece, discrepancies in the number of parts or in the times required to deal with them and so on. In other words, those difficulties that were originally the job of the human being to overcome.

"Artificial intelligence" is to endow machines and their computer-controlled co-ordination, with that capacity for adaptation which they previously lacked. "The important things here is the similarity between their performance and the intelligent forms of behaviour of the comparable operating personnel" (Spur 1984, p. VI).

This similarity is really no surprise if we consider the way that work has been divided up, subjected to machine operation and expressed in rules. The scandal lies in simply taking those particular human tasks that have been reduced to machine-like operation, and then using them as a measure of human intelligence. The illusion, which becomes grander the further removed from practice one becomes, is to believe that it is possible to replace the last outposts of human labour with expert systems, and still be able to maintain adequate adaptability in production systems. The depths of error into which it is possible to sink as a result of this machine-like thinking, may be illustrated by the production planner who considered the use of an expert system to be essential because the production process in question was characterized by such complex interrelations, displaying unknown and unpredictable factors, that human beings were being overtaxed in planning and controlling it. Statements like this (by no means an isolated case) show that in the misguided attempt to overcome their natural stupidity by means of artificial intelligence, many scientists and engineers have literally, lost their reason.

The kind of illusion and blindness evident in this way of thinking, is in no way a matter of chance, but is characteristic of AI researchers and their disciples. They need and deserve to be scrutinized in a critical way, in order that the origins of such an erroneous viewpoint can be traced.

Just recently, two eloquent supporters of AI research, yet again formulated claims and promises which quite simply leave one speechless and challenge one's reason. These were made in a bombastic propaganda publication on fifth generation computers, and were in complete agreement with other leaders of their fraternity. In this publication, they announce to an astonished world a "new wealth of nations" (Feigenbaum and McCorduck 1983, pp. 5 and 15ff), and they name the basis on which it is to unfold: "the reasoning animal has finally made the reasoning machine" (ibid. p. 7). They dismiss any serious doubts about whether machines could be capable of thinking at all by merely listing them without even the slightest attempt at questioning them. This is done in order to be able to define away the whole question of whether a machine can think as "a non-question, a non-issue of no consequence" (ibid. p. 45). In so doing, they evade the real question in dispute, namely, what intelligence actually is, by simply reverting to Turing and attesting that a machine exhibits intelligent behaviour if it can solve a problem which, according to popular opinion, demands "intelligence" equal to or better than that of a person ("a machine as smart as a person", ibid. pp. 40ff). But what if, contrary to popular understanding, this problem called for a well-defined procedure and therefore no particular insight at all, or if the comparable person was not sufficiently experienced for the task?

Feigenbaum and McCorduck have thus turned out to be genuine representatives of their fraternity, which has always allowed itself to get so carried away with its bold assertions that it omits to put them to the test. For example, in 1957 H.H. Simon predicted that within ten years, a digital computer would be chess champion of the world and would discover and prove a new mathematical theorem. He said as well, that most psychological theories would be available as programs (Dreyfus 1979, pp. 81ff). What bad luck for the prophet that so far, none of his prophecies have come true!

In all the principal areas of AI research, development has followed the same pattern. After amazing success at the initial stage, nothing more is discovered and any insight after that, is dissolved in failures. This is what has occurred in (say) the automatic translation of natural languages; in general problem solving; in the identification of models and in chess. Unfortunately, there isn't room here to subject the individual research efforts of AI to the criticism they deserve. (For a more detailed expose I would refer you to the extremely sound and perceptive book by H.I. Dreyfus, "What Computers Can't Do".)

This course of events has not remained totally unnoticed by the protagonists and so they are formulating their dreams with a little more restraint:

Feigenbaum and Feldman claim that tangible progress is indeed being made, and they define progress very carefully as "displacement toward the ultimate goal" [namely, the creation of a thinking machine]. According to this definition, the first man to climb a tree could claim tangible progress toward reaching the moon. (Dreyfus 1979, p. 100)

This stubborn adherence to the aim of programming a digital computer which could outdo a person in terms of intelligence, is comparable to that of the alchemists, whose unshakeable efforts to make pure gold from other substances is an object of ridicule today. Nevertheless, this programming goal is still being pursued, even though the results so far achieved are anything but encouraging. The unbridgeable gulf between what these people want to do and what can actually be done in practice, deserves to be explained in greater detail.

It is thanks to his hands that a person can think, and thanks to his brain that he can work. Right from the outset, work has been peculiar to the human being as a practical and expedient activity for producing food to satisfy his needs. In active interaction with nature (the reality existing independent from himself), and in co-operation with other individuals (which forms the basis for social relationships), he gathers experiences from his intervention in the natural sequence of events. He then perceives things that can be made use of, in order to act more purposefully upon this same sequence and, by grasping and deploying them, he understands their function and recognizes the way they work. In short, he forms a conceptual model of the process of interaction with nature. The components of these models, called "schemas" in psychology and which consist of things that have been understood, such as experiences of objects perceived and activities performed, are generalized and objectivized for further use as tools, and in the form of language. Through the use of language, people can agree on joint and purposeful action, and can pass on to future generations, the accumulated experience gained from these actions, and from the use of tools (Holzkamp 1978, pp. 147ff; Volpert 1984a, pp. 81ff).

Thinking is nothing other than trial behaviour going on in the human mind. The actions, whether they are implemented or simply conceived, are not usually stored as complete models of a sequence of events, but are generated when required by drawing on the schemas. We explained earlier how, using the division of labour, elementary actions with strictly predetermined sequential forms (so-called automatisms) can be created and expressed as algorithms. Machines are their objectivization. Since objectivized schemas such as concepts, tools and machines make certain demands on the human acting, any lasting and pervasive interaction with machines, leads to the withering away of human activity into a machine-like routine, resulting in the mechanization of both thought and action. We can see then, how it is possible to automate intellectual work.

Human action is carried out through the interaction of hand and brain and is driven essentially by the internal motives of the relatively autonomous subject (the person themselves), not by external specifications. The action is constantly embedded in a situation. It comes about because of the combination of the physical and mental requirements of the acting subject, his experiences in the form of schemas (from which he constructs the internal model of reality), the concepts and tools handed down and the external connections between nature and the society surrounding him.

Now, when action is performed in a situation, the body plays an important role. We use it to transform a need into an impetus and an objective for action, and its senses are the basis of perception, although this perception is strongly influenced by the internal model. Only by using his body, with its highly developed motor activity, sensitivity and perceptiveness does the human being succeed in forming schemas from his experiences. As a result he develops expectations of a situation (anticipated perception, the "inner horizon") and reveals the hidden aspects of the situation by means of which he recognizes the contrast between figures and their background, and between the essential and the incidental. The latter problem is one which AI research has not even attempted to resolve:

The body contributes three functions not present, and not as yet conceived in digital computer programs: (1) the inner horizon, that is, the partially indeterminate, predelineated anticipation of partially indeterminate data (this does not mean the anticipation of some completely determinate alternatives, or the anticipation of completely unspecified alternatives, which would be the only possible digital implementation); (2) the global character of this anticipation which determines the meaning of the details it assimilates and is determined by them; (3) the transferability of this anticipation from one sense modality and one organ of action to another. All these are included in the general human ability to acquire bodily skills. (Dreyfus 1979, p. 255)

The important thing about skills is that, although *science* requires that the skilled performance be *described* according to rules, these rules need in no way be *involved* in producing the performance. (ibid. p. 253)

All this can be summarized in two theses, which are mirror images of one another. One is that human beings are able to think thanks to their hands (which explains how intellectual work can be automated within limits) and the other is that since a computer has no body, it cannot think (which explains the limitations of AI). We can thus define a person's intelligent behaviour as his ability to act purposefully in a situation of uncertainty, even without established rules.

Human intelligence can thus be identified as the exact opposite of what it appears to be in the minds of those who have been blinded through thinking in machine terms. It is the active use of self-will in unknown terrain, rather than passive adaptation to changing circumstances. It is the upright, autonomous subject creating his own circumstances, not the pressure of circumstances on a so-called "intelligent" object which has to adapt itself to them. Nowhere is the fetishism of thinking in machine terms expressed so blatantly as in the behaviouristic view of intelligence, which cannot perceive human behaviour as anything other than a mechanical analogue.

The limitations of AI are not only due to the fact that a computer is not capable of thinking because it has no body, is not involved in a situation and has no intentions. Conversely, a human being, because his actions are linked to intentions and are bound to situations, is unable to program the computer to compensate for its deficiencies. He is unable to ensure that right at the beginning, an evaluation of all possible situations is contained in the program. It is also impossible to think of a complete set of rules that could be drawn up in advance to achieve this: That the computer could generate behaviour appropriate to the situation in all circumstances, on the basis of data gathered from the surroundings.

These limitations in human thought are caused by the fact that the human ability for abstraction is limited to discovering the general only from that which has already been experienced, recognized and handed down. Therefore, anything that has been thought out in advance always contains the roots of its birth – the situation and the intention.

This also explains why it is that AI research has recorded partial successes in relatively well-structured fields of activity, where heuristic rules can be sensibly applied, as in chess for example. It has failed everywhere in fields of activity which are not well-structured. This is because events often have an implicit meaning which is evident only from the situation itself (as in understanding natural languages; Dreyfus 1979, p. 292). To recover from such failures, many

AI researchers seem to have abandoned the search for general principles in the programmed intelligent behaviour of computers. It seems as if practically the whole fraternity has reverted to the development of knowledge bases and expert systems and into areas where some successful applications are to be expected. Already by 1984 about 140 expert systems were being developed or tested worldwide, mainly in the fields of medicine, chemistry, the design and configuration of computers, production planning, error diagnosis, terrestrial exploration and the law.

Without any proper results being available, "knowledge" is already being launched, amid a great deal of noise, as the fourth factor in production. Since except for the first factor (labour), the second (capital) and the third (land) turned out to be a fantasy of middle class ideology, we may be forgiven for eagerly wanting to know what the fourth (knowledge) has to offer.

Expert systems promise to place more knowledge and more power in the hands of the manager working in the service of capital, which is now in crisis everywhere. With the fifth generation computers, and massive use of knowledge bases, the manager is given to understand that he will have at his disposal all the knowledge required for production, without large numbers of expensive and self-willed experts. In this way, he would control production and be able to adapt it where necessary.

This however, is precisely the illusion which appears in the minds of management personnel as a result of viewing knowledge as an independent factor in production. For as we have shown, knowledge is primarily nothing more than experience that has been converted in the human mind into schemas and then used to form conceptions of intended action. It is only by some form of objectivization that knowledge can be given tangible form in language, tools, organizational forms, machines and literature. It is true that these demand further action, and that they determine the way the actions are carried out, but to have a useful effect they must be sensibly used by human labour. Thus there can be no talk of an independent production factor, except as a fantasy in the minds of ideologues.

Not even the knowledge base itself can be the source of further knowledge, although this is exuberantly proclaimed by Feigenbaum and McCorduck in the style of the fraternity: "The computer *produces* information. The essence of the computer revolution is that the burden of producing the future knowledge of the world will be transferred from human heads to machine artifacts" (Feigenbaum and McCorduck 1983, p. 40).

Despite this belief in miracles, it is not old knowledge that creates new knowledge but lack of knowledge that is a necessary precondition for understanding, and lack of understanding that is the stimulus for making use of and going beyond, existing knowledge to obtain new knowledge by means of action. Without a conflict between available knowledge and reality that the human mind cannot yet resolve, there is no discovery. At the same time, this procedure for understanding is the reason why knowledge previously acquired becomes obsolete and superfluous when it is replaced by more recent findings. For its owner, knowledge is a highly perishable commodity.

Ironically, the technology of knowledge engineering leads to a contradiction. The more "artificial intelligence" these expert systems contain, the more they require clever human minds to interact with them in order for them to be effective. Conversely, the more widely these expert systems are used to replace human experts, the greater the loss of ability to discover anything new. Once again, management is increasingly caught up in the contradiction between the expediency of cheaper, more reliable and more independent expert knowledge, and the necessity of having to adapt to new and changing circumstances.

Difficulties proliferate in another area however, that of eliciting expert knowledge from the human mind in the first place, and giving it a suitable tangible form as a knowledge base (knowledge acquisition). Ironically, the "knowledge engineer" whose task it is to cope with this dubious activity, lacks the relevant expert knowledge himself, but is nonetheless expected to elicit rules from those who do possess it: rules that these experts are not in fact aware of themselves. It is no lack of goodwill on the part of experts that prevents them from providing knowledge. It is the fact that they cannot call upon this knowledge in the form of explicit rules. This is a fundamental obstacle in getting at this "tacit knowledge", knowledge which is indebted to the functions of the body during the thinking process, and it cannot be fundamentally overcome.

These difficulties are recognized in AI research circles too of course, but from a perspective that has been distorted by thinking in machine terms. Feigenbaum and McCorduck, who can only envisage the knowledgeable human brain as a pit to be mined, write thus: "The heuristic knowledge is hardest to get at because experts rarely have the self-awareness to recognize what it is. So it must be mined out of their heads painstakingly, one jewel at a time. The miners are called '*knowledge engineers*'" (Feigenbaum and McCorduck 1983, p. 77).

Difficulties accumulate elsewhere when we examine the question of whether a knowledge-based system will behave as anticipated. The technical problems of accuracy, consistency and reliability are as yet, completely unsolved. Heuristic rules, deduced from the observation of human experts and in consultation with them, often turn out to be incorrect, inconsistent and unreliable. Yet, whilst human beings can cope with such vague notions, due to background knowledge of the world in which they live, knowledge-based systems cannot, and this leads to serious deficiencies. For in each case arising, where a solution cannot be found using the knowledge already at the system's disposal, new rules have to be added. So the inconsistency tends to grow further, and the harder it becomes to comprehend and predict the behaviour of such a system, the more untenable it becomes to attempt to verify solutions already found at an earlier stage. Maintenance becomes a nightmare.

Due to its contextual dependency, it is not possible to define precisely, the limits of a knowledge base. Consequently, the system itself cannot determine whether or not it is exceeding its own competence. Even the so-called "explanation components" implemented in many systems (which are in effect, simply the sequence of rules used for inference), are to little avail. Basically, knowledge-based systems deserve even less confidence than those "spaghetti heaps" of unstructured conventional programs, whose unreliability and confusion were such a bone of contention in the development of modern computer software (Parnas 1985).

Even greater problems arise with regard to the assimilation of human expertise by these systems. As described above, new insights and new knowledge can only be acquired by human experts. The assimilation of this expertise is therefore absolutely imperative. Meanwhile, close analysis reveals how unique human capabilities find expression in specialist expertise. Dreyfus and Dreyfus define five stages of expertise. These are summarized in Table 2 (Dreyfus and Dreyfus 1986).

Table 2.   Five stages in skills acquisition (from Dreyfus and Dreyfus 1986, p. 80)

Stage	Components	Perspectives	Decision-making	Attitude
1. Novice	Context-free	None	Analytical	Dissociative
2. Advanced beginner	Context-free and situational	None	Analytical	Dissociative
3. Competence	Context-free and situational	Selective	Analytical	Dissociative understanding and decisions, emotionally involved in outcome
4. Skill	Context-free and situational	Experienced	Analytical	Involved understanding, dissociative decisions
5. Expert	Context-free and situational	Experienced	Intuitive	Emotionally involved

When people enter a specialist field, they use explicit book knowledge only up to the advanced beginner stage. After that, they proceed in accordance with known, objective and context-free rules comparable to those implemented in knowledge-based systems. Crucial to the higher stages of expertise in unstructured problem areas is the fact that at first (stage 3), they acquire the ability to perceive a situation without any explicit rules, and to select appropriate target-oriented strategies. This ability is only attainable through experience, and on occasion, can even involve breaking the rules. Only at stages 4 and 5 can we discern the ability which finally makes them capable of intuitive action and of recognizing crucial similarities between situations. The ability is attained only by acquiring further experience, and by virtue of being emotionally involved and actively caught up in the world of problem solving.

We can clearly understand from the above analysis, how it is, and why it is, that on the basis of their design criteria, knowledge-based systems can at best only reach the performance level of the advanced beginner. If knowledge-based systems were to be used today, in a specialized field of knowledge, then the knowledge and skill of human experts would not at first be required. The less they are called upon to solve problems, the less frequently will they be presented with a challenge. As a result, their capability will gradually diminish. That ability to make assessments and adapt to unspecified situations will wither away, and this is precisely the attribute so urgently needed for solving unstructured problems and updating existing knowledge.

On the other hand, the blithe assurance that knowledge-based systems would not replace human experts but merely support them and complement their expertise, is heard again and again. In the light of the way in which the interaction between systems and experts is organized, and the way various functions are allocated, this can only be taken as empty rhetoric. In order for people to develop their unique abilities, as expressed at the highest levels of expert activity, they must be in a position to experience the effects of their activity, and this has to happen in the context of the preconceptions that were implicit in their intentions. It is precisely this however, that is prevented in the current concept of knowledge-based systems. The system responds on the basis of the context-free knowledge available, and not in the light of its appropriateness, not even with the aid of "explanation components".

Experts forced to work with these systems are therefore in the unhappy situation of doubting the correctness of the system's responses, while at the same time being unable to make any check. Any expert would be condemned to losing his real expertise over a period of time, and slide back to the competence level of the beginner. The more that "artificial intelligence" is incorporated into knowledge-based systems, the more it needs the interaction with human minds of yet higher capability in order to be able to use that intelligence. Conversely, the ability to find out new things would be lost even quicker should these systems become more widespread as a replacement for human expertise. So any management would increasingly find itself in a catch-22 situation, on the one hand striving for cheaper, more reliable and more independent expert knowledge, and on the other hand needing to adapt repeatedly to new sets of circumstances.

All this begs the question of how knowledge-based systems should be designed and it is crying out for a new answer (Winograd and Flores 1986). It is no longer enough to go prospecting for the knowledge of experts and then objectivize it all in a system. A theoretical framework must be provided instead, on the basis of the reality of human behaviour, which makes it possible to work out and to understand the unique nature of human thought and activity as opposed to that of machines. In this way, there could be a human-related interaction and function sharing between human and machine ("contrastive work analysis", Volpert 1987).

Approaches based on cognitive science and "artificial intelligence" will not do here, because they seek to explain human behaviour by modelling it on machine behaviour. Secondly, the development of the system must no longer be mistakenly seen as a purely technical exercise, but has to be perceived as a social relationship between the developer of the system and the user, whereby the developer sets the operating conditions for the user. The situation at present is that every technical method of systems design is too simplistic, for it does not encompass a comprehension of social interaction or of human activity. Thirdly, systems development cannot be successful without shaping the work process in question at the same time; that is, tailoring the various tasks, the function sharing and the human–machine interaction. Since this design task cannot be objectivized, and since it relies largely on social interaction, the only design method that that can be considered is a participative and interdisciplinary procedure in which right from the outset, design is subjected to the discourse of management, systems developers and systems users.

Bearing all this in mind, we are able to describe the problem areas in which the successful use of knowledge-based systems might be expected, and those where the preconditions for success are not satisfied. The appropriate criteria are met everywhere; they are:

1. The required knowledge can be objectivized.
2. A limited number of objects are to be represented with their characteristics and relationships.
3. There is precise knowledge of the problems and the methods for their solution.
4. The competence limits of the system can be explicitly ascertained.

A typical class of problem in this category is that of systems configuration because the available modules, their compatibility conditions, and the tasks can all be formally described without presenting the system with insoluble problems.

It is no surprise therefore, that the XCON expert system for the configuration of VAX systems was a widely quoted success.

It would appear that diagnosis is already a problematic area. The multiplicity of developmental and operational experiments going on in the design of systems for diagnosis is not a consequence of its special suitability, but a sign of the difficulties laid bare. Only when the set of symptoms, the set of causes of failure and the relationship between these two phenomena can be stated clearly and in a context-free manner, will there be any chance of success. As soon as it becomes necessary to have recourse to implicit background knowledge (as is usually the case in medicine), problems such as those described above will arise (Puppe 1987).

The situation is no better with the automatic interpretation of natural speech, which has been seen as a possible target area when elements and relationships can be expressed explicitly, but which turns out to be a hopeless case when its contextual dependence cannot be made explicit (Winograd and Flores 1987).

The point has now been reached where we must search for the origin of machine-like thinking. How do we explain why intellectuals, whose intellectual work is in no way subject to a mechanical mode of operation, can only think of human labour as being mechanized? Such ideas can be traced back into antiquity via a long chain of Western traditions of thought. Among the first philosophers, Plato put forward the idea that "knowledge" deserves its name only if it can be expressed in terms of explicit definitions (leaving only the meaning of fundamental concepts to intuition). But why did such ideas from the ancient world of mythology suddenly spring up just at this time, and in this particular place?

Plato's own thoughts give us some information on this. He strictly differentiates between that which is in the process of becoming something else and never remains the same (which people comprehend with their senses), and that which never changes, which just exists (which people comprehend in thought, in ideas). These two are related in a participatory manner, in just the same way as in the linguistic form of a sentence whereby the noun names an object and expresses what it is in the predicate, that is, it gives it meaning. Now the relationship described is of exactly the same type as that in respect of goods. The use value of goods (like sensory impressions) are as varied as there are purposes for their use. Consequently, they are incomparable and yet (just like ideas) they have something in common which is the equivalent for which they are exchanged, the exchange value. Every article is unique as far as its use value is concerned, and yet by means of exchange, it is made equal to other articles. It is desired by someone else precisely because it is unique, but because it is exchanged for another article it is made equal to it. "Goods have no use value for those who own them. They do have use value for those who do not own them". These people "can relate their goods only in terms of a value and therefore their articles have to be compared by relating them to some other article which is a general equivalent", as Marx was able to say with his own brand of clarity! (Marx 1969, pp. 100f).

This abstraction of the exchange value from the use value does not take place in people's minds but in exchange itself, in a very real sense, that is, by goods being exchanged with each other through the act of exchange, through the actions of people, but without making use of their knowledge. "The exchange of goods is abstract, because it is not only different from their use, but is also separated from it in time. The act of exchange and the act of use are mutually exclusive in time" (Sohn-Rethel 1973, p. 47). In this process, it is merely the ownership of the goods that changes, whereas the use value cannot undergo any changes at all (compare this with the idea of pure space/time motion). This leads to a further abstraction:

The goods are not equal; exchange makes them equal. This equating performs the abstraction of the quantities of goods available for exchange into abstract quantities only. The exchange equation eliminates those determinations of amount that are related to the utility value and cannot be compared with each other. It replaces these named quantities with an unnamed one, which is nothing more than simply quantity, unrelated to any kind of quality [compare this with the idea of the continuum of numbers]. Like the exchange equation from which it sprang, this quantity, in itself or in the abstract, is of a relational nature and, like the exchange equation, adheres in turn to the act of implementation. (ibid. pp. 74f)

So it is not the consciousness of those performing the exchange that is abstract, only their action. Since both are necessary, the abstract nature of the action and the non-abstract nature of the consciousness that accompanies it, those who are performing the exchange are not aware of the abstract nature of their action". (ibid. p. 49)

It was only in the three centuries preceding the death of Plato, that the trading of goods rapidly developed from a sporadic simple exchange activity to a universal social mode of trade, bringing with it far-reaching, radical social change. It isn't surprising in the circumstances, that this should occur around the Mediterranean, where the sea, unlike the land, provides a suitable trading link. Only in its developed form, was the trade in goods able to re-mould the abstraction taking place in the act of exchange into a general social relationship, thus transforming this abstraction in the human mind into the ideas of "pure reason". In this developed form of trade, the value of the article became independent through money. This is indicated by the coinages proliferating at this time, being the tangible material bearers of the general equivalent (Thomson 1961, in particular pp. 159f, 253f and 288f).

The abstraction of exchange is not thought. This fact provides us with the key to understanding the historical genesis of "pure reason" from the social entity. The elements of the abstraction of exchange are reflected, given the right social conditions, as pure concepts in the consciousness of those who possess money, because they are pure abstractions contained in the social entity. (Sohn-Rethel 1973, p. 99)

We have shown here what Platonic teaching was based on, as the origin of this thinking in machine terms, and how it could originate out of pure thought without any reference to physical work. We have also explained why the separation of hand and brain has to lead to the mistaken belief that it is possible to open up all knowledge through "pure thought" with explicit definitions and rules, and why the functions of the body which are involved in intelligent behaviour do not need to be understood. Thinking in machine-like terms is thus revealed as a particular type of goods fetishism, by virtue of which the social relationships of their intellectual work, appear to those who are thinking in machine terms "as the objective relationship of people and the social relationship of things".

However, thinking in machine-like terms would not have the true character of a fetish, if it didn't make use of all kinds of mystification and "theological overtures". When we have just got used to the idea that human labour is being ousted by "intelligent" machine systems and (inasmuch as it is still needed at all), is being subjected to the functional requirements of these systems, there appears suddenly and quite unexpectedly from these systems a *deus ex machina*:

"In this new technology, human creativity becomes the motor which drives and controls all developments and which, like a narrow pass, determines the speed of the process. The engineer will need the creativity of the artist. Technology and art, work and play are converging" (Spur 1983, p. 25).

3.4.3   Deskilling and its Consequences

The technocentric concept of production is the result of a specific set of social driving forces, set on rationalization and seeking to overcome current difficulties through automation. It is characterized by:

1. The demands resulting from increased competition, i.e. to produce even faster, better and more cheaply.
2. The need to maintain dominance and continually counteract the self-will of capable working people, which entails increasing the transparency of work and management's control over the production process.
3. Thinking in machine terms only, and thus viewing machines as a panacea for displacing human labour, which is considered unreliable, inefficient and a source of problems rather than a productive force.

It is the skilled work still being done in the workshop and the intellectual work carried out on an increasing scale in the technical office, that offer the greatest potential for such rationalization. But the highly complicated production process itself, the variety of forms, the complexity of its products and the flexibility demanded by the market all militate against comprehensive rationalization. It is logical that rationalization measures should begin in those areas that are most easily adapted to computerization as the universal vehicle for rationalization, and where the effects of it are the most promising. The knowledge and abilities of skilled workers in the workshop and specialists in the technical office are given concrete shape in formalized sequences, databases and programs. To that extent, they see themselves being driven back into dealing with residual functions in which their capabilities are wasted.

The responses provided by a technocentric development strategy to the challenges of the present position are extremely questionable. They are reminiscent of the race between the hare and the tortoise, for every time an issue is settled, it reappears in a new form. The basic problem to be solved in this strategy is to bridge the gap between the analytical model of production and the actual course of events. To this end, it is demonstrated over and over again, that concessions have to be made to human labour in respect of its authority to act. These concessions however, are quite incompatible with the long term avowed aims of the technocentric production concept.

In addition, the opportunity to provide simple and consistent external factors that affect production, so that planning and implementation can coincide, is lost in conflicting market demands. Consequently, the only hope lies in first assembling all functional production areas by means of an integrated computer system with a common database, and operating them as a combined whole. Even if success were to be achieved in giving an objective form in the computer system, to all the knowledge essential for automated production, the clumsiness of the programmed sequences would make production too inflexible. In this situation, the more intense use of expert systems offers a promise of help in avoiding the threatened rigidity. But the limitations of "artificial intelligence" restrict the effective use of expert systems to those problem areas and situations that are relatively well structured and display regular behaviour in a manner that promises success. The situations that most frequently arise in the capital goods market, cannot be mastered with expert systems, for they require new types of flexible procedures, free from predetermined rules.

At every successful stage of the journey along the route of technocentric development, human labour would be driven further into the nooks and crannies of the residual functions not yet automated, and become subject to more and more conditions set by machines. At the end of the road there would be the integrated computer system with all the functional areas of production brought together, along with its inevitable counterpart the labour force, among these functional areas, and performing the disconnected residual activities. As subjects, human beings would be excluded from the role of actively shaping this part of their lives. Inasmuch as they would still be involved at all, they would be subject to conditions of work which would cause their capabilities to wither away for lack of practice and their creativity to disappear, resulting in their own machine-like behaviour.

Of course, this development route leads through rough terrain, where obstructions pile up and can only be overcome with difficulty, if at all. For one thing, substantial parts of the production do not lend themselves to analysis and algorithmic description, either because they are too complex or because they form entities that are not susceptible to being divided up into elements and rules. More often than not, the experience of human experts is unavailable in concrete form, because they have this experience available to them as "tacit knowledge", not in the form of explicit rules. But even where the analytical formulation of models is successful, the complexity and the flexibility required in production make the development and (even more so), the maintenance of programs so expensive, that capital intensity itself grows at a disproportionately high rate.

The high investments and risks involved in all this, come into direct conflict with the restricted finances of the capital goods industry, which is made up in the main, of medium-sized enterprises. Such development could thus only take place in a few large enterprises, and the majority would be bypassed. Small and medium-sized enterprises however, are of particular economic importance. Contrary to popular opinion, mechanical engineering firms in Germany, with 20 to 299 employees were able, between 1971 and 1980, to increase their share of work from 23.2% to 29.3% and their share of turnover from 21.4% to 25.5%. In addition, companies seem to be restricting themselves more and more to the strategically important core parts of the production process; those which enable them to control lead times and ensure continuity of knowledge. They are handing over the production of the remaining parts to a large number of foreign firms (VDMA 1982; Manske and Wobbe-Ohlenburg 1985). One would expect this to throw up unforeseeable difficulties concerning competitiveness, in view of the many-faceted and still growing number of exchange deliveries taking place with these foreign firms.

Because of the concentration of knowledge in the technical office and in the integrated computer system, all orders must first of all go through the planning process before they are released for production. As before, this is organized on the principle of performing routine tasks. Although the use of computers increases the clarity and speed of events, the two main causes of long throughput times and fierce resistance to change, nevertheless remain in place: (a) the principle of functionally divided workshops, and (b) the initial central planning process. Hence production comes into conflict with market demands for greater flexibility in product modification, quantity and delivery times.

In the case of changes in the production process, the new means of production and procedural sequences must first be represented in detail in the computer system. However, this design of computer models requires very demanding and complex analysis and testing, since essential production knowledge has been removed from the residual human labour, and its capabilities have wasted away. Innovation in production is becoming extremely complex, and alongside that there is a growing danger that the capacity for innovation will be gradually lost. Creativity is fast disappearing, even in design, so too is the ability to improve on product design or to design something new. Thus, those abilities which are in danger of being undermined, are precisely the ones on which competitiveness most critically depends.

It has to be expected, that despite complex planning and accurate model design, the susceptibility of the machine system to faults and disturbance in unknown situations, will increase as a consequence of integration and central-ization. Even so, high capital intensity still requires total, continual availability. Overall, the implementation of the technocentric production concept leads to a situation where the very considerable skills and capabilities of the workshop and technical personnel, are no longer needed, and so they gradually disappear. At the same time, this development requires new capabilities that are not available.

Computer-integrated manufacturing is run by a small number of highly trained specialists who have to be prepared for their task in a similar way to an aircraft crew. They have overall responsibility for the correct functioning of a complex technological system, and they are assisted by a group of specialists whose job it is to eradicate errors and carry out rapid repairs. (Spur 1983, p. 24)

So the costly and somewhat dubious development of the "workerless factory", originally conceived as being independent from human labour, ends up in a situation where computer integrated manufacturing still needs the capable work of specialists in order to function smoothly. So, automation reveals its ironic aspects yet again (Bainbridge 1982):

The more automated a system, the more important is the human contribution to its operation.

The operators are trained to obey instructions but they are still expected to make an intelligent response in the event of an irregularity.

A system is considered to be particularly successfully automated if it requires only occasional intervention by people, who nevertheless have to be highly trained.

Chapter 4

The Anthropocentric Route: The Return of the Human Being

Anthropocentric production provides a fundamental and universal alternative for production engineering design and labour organization. This means, among other things, that it can incorporate different levels of technological and organizational development and does not function simply as a stop-gap for tackling the unresolved contradictions of the "workerless factory". This development can be pursued under conditions of competition, and, as will be seen, to greater economic effect than technocentric production, that is, when the fundamental principles are demonstrated to full advantage. For the purpose of greater clarity and deeper understanding, the basic properties and features of the anthropocentric concept will be compared with those of the technocentric one using idealized models.

In doing this, we must not of course overlook the fact that the anthropocentric route can only be established as a branch of the technological one, for it has no independent starting point. However, the contradictions of the "workerless factory" and its obstructions to this new course of development present a challenge for the new route being carved out. The transfer from one production concept to another may seem like a qualitative leap, a fundamental change in the perception of the human/machine interaction. It can however, be carried out at any time, since anthropocentric development can occur in stages. This is also true of course for the reverse process, that of returning to technocentric production.

In the areas where these two opposing concepts are in contradiction, the outcome of rationalization has so far, turned out to be extremely unstable. The actual path taken toward rationalization from that point, depends on market requirements and on the production economy itself. It depends too on the social forces a business is subject to, the internal conditions for action, and on how far the protagonists have freed themselves from the constraints of thinking like machines.

The following preliminary remarks are of importance in order to avoid any misunderstanding. Firstly, the straightforward, idealized depiction of anthropocentric production, with a description of its essential features, should not blind us to the fact that there is an endless variety of possible intermediate forms of production technology and labour organization. These contain features which are characteristic of both concepts, the technocentric one and the anthropocentric one as well. Furthermore, it is necessary to identify the obstacles placed in the way of full development of the anthropocentric concept. In principle, that concept encourages the design of forms of production technology and labour organization which are more suitable for people. However, so long as the demands of competition exist and decisions are made externally, the imperatives of capital deployment and workplace control will still retain their significance, even when human-centred forms of production are envisaged.

In following either of these courses of development, the more or less intensive use of computers is not an issue, neither is the higher or lower degree of automation. It is rather a question of the way people and machines work together bearing in mind the vastly differing developments and applications in production technology. Instead of emphasizing the potential of the technology itself and stressing the limitations of people, the anthropocentric production concept seeks to develop the creativity and productive capabilities of human beings to a far greater extent through their work with machines. The machines must therefore be designed to fit in with this concept, and the whole process will be completely different from machine design in technocentric development. Instead of imitating human abilities and reducing the human being himself to functioning in the same way as the machine or indeed, displacing him from the production process altogether, the intention is to combine the complementary attributes of human being and machine in a productive way.

4.1   The Workerless Factory: Signs of Return

4.1.1   The Dilemma of Radical Change

The practical effects of technocentric production, dealt with in earlier chapters, have brought to light a series of contradictions which cannot be solved within the technocentric framework itself. On the contrary, the contradictions become more acute as this development proceeds.

1. The more frequently management uses this concept to try and displace human labour from the production process, the more precisely and completely must its analytical model of the individual procedures be worked out.

Once the routines are in place, subordinates are meant only to apply them. Deviations from the routines are more frequent and necessary to the organization's success than management thinks. Rapid changes in production goals are at odds with this low trust system. The more frequently products and processes are changed, the less time there is to translate conceptions into reliable, mechanically applicable routines. The more imperfect the routines, of course, the more interpretation and initiative they require from workers at all levels. (Sabel 1982, p. 210)

The deviations generate a desire for even more accurate planning, and this further increases the complexity of the process, which is already complex to start with. This planning is removed still further from implementation so that in the end, the gulf between model and process is widened still further. As a result of all this, management gets caught up in the first of its dilemmas. *Does it become trapped in the vicious circle of ever more comprehensive planning, with ever increasing deviations from these plans or does it, in contrast to its original intentions, draw up only rough planning specifications which leave room for manoeuvre on the part of those implementing them (for example, batch control or the reintroduction of work planning into workshop production)?*

2. Closely related to this is management's aim of increasing the clarity and control of the work process. Every step which removes the knowledge required from those performing the work and causes it to be objectivized in a machine, can provoke subjective resistance on the part of these operators. In addition, it creates objective conditions whereby their expertise, frequently drawn upon in unforeseen circumstances, just fades away.

Workers in low trust organizations are presumed not to share the organization's goals. They are neither trained to show, nor rewarded for initiative. Hence they are likely to regard the initiative that they do have to exercise to keep things moving simply as a bargaining chip in their struggles with superiors. The classic example is the strategy of working to rule: stopping production by going by the book. (ibid. p. 210)

Unfortunately, it thus comes about that maintaining control in the factory, destroys the conditions under which human labour might develop productively, even though its expert knowledge is called upon over and over again. Management thus gets caught in the second dilemma. *Should it push through its bid for control and risk lowering productivity levels, or should it go against its own wishes and grant those involved in the plan the authority to act on their own initiative (for example, allow workshop programming)?*

3. Capital deployment in changing market conditions, places big demands on the production process. These demands are hardly compatible with central planning and routine working, which has arisen out of the horizontal and vertical division of labour. This division actually results in limited flexibility, longer throughput times and delayed delivery dates. Every measure which deploys computers within this structure, for the purpose of reducing throughput times and hence costs, will sharpen the structural conditions that restrict any such outcome. In order to satisfy market demands to the extent required by capital, the production structure would have to be designed for flexibility by reducing the division of labour. This in turn would run counter to the management's need to maintain dominance in the workplace. Management thus becomes caught up in the third dilemma. *Does it save time and costs and thus risk a reduction in flexibility or, contrary to its wishes, does it reduce the division of labour?*

4. Taken together, these three dilemmas form the basic contradiction within technocentric production. The market demands a flexible production process with short throughput times, but the structure of a process organized on the basis of the division of labour militates against this. Conversely, the demand for transparency and control is satisfied by precisely that form of division of labour which obstructs production economies and market demand. Development has thus reached a point where the requisites for maintaining control are in conflict with those for deploying capital. Resolving this conflict, or at least taking the edge off it, will have to be the result of a quite different production concept.

4.1.2 Management Split

The contradictions in technocentric production have not of course gone unnoticed by its protagonists. They are clearly reflected in a fierce debate about where the solution lies in this situation of radical change. It is within management, where the majority seem to think along Taylorist lines, that the march towards the "workerless factory" is being subjected to growing challenges both in quality and quantity. It has been possible more recently, following arguments over the workshop programming of CNC machines, to follow the public debate as well, which sometimes degenerates into a battle over dogma (this is not surprising, considering past history). The debate is not about marginal technological or organizational alternatives for this or that production task. More far reaching than that, it is about the correct long term concept for the restructuring of the whole production system. Advocates of a new type of production concept are logically consistent in making their springboard for argument, the central weaknesses of the concept they are attempting to overcome.

In the Federal Republic of Germany, the Committee for Economic Production (Ausschuss für Wirtschaftliche Fertigung – AWF), has formulated the concept of the "production island" as the basis of an alternative production structure. The structure was worked out in collaboration with a number of important mechanical engineering firms over many years of intensive work in its "production islands" study group. It has all been summarized in a handbook (AWF 1984), in which the change in the nature of demand and the unsatisfactory labour situation are taken as the starting points for a fundamentally new production concept:

From this point, a number of demands are made on future production: rapid adaption to current turnover; quick, flexible reaction to qualitative changes in demand; high flexibility in dealing with the effects of technical faults combined with minimal downtime; short throughput times in tandem with fewer machine link-ups, so that they can perform a variety of functions; high quality, achieved with economy; the introduction of new products or variants into current production, in the most problem-free way, as well as their problem-free changeover. (ibid. p. 1)

The labour situation, and the attitude of employees to their work are considered decisive and at the same time, somewhat awkward:

In this connection, a disturbing tendency has arisen in the last few years which is evident from the regrouping of the workforce in production workshops. The departure of a large number of dissatisfied workers was not considered sufficient reason to address this tendency by redesigning production. Advanced production engineering of the future will increasingly require better qualified workers who will be prepared to accept the current conditions of production to a limited extent. This can lead to the negative aspects of traditional working conditions exerting an influence on the attitude of the workforce toward the enterprise. Fluctuation in performance, absenteeism, increased disruption and a high reject rate, all symptoms of dissatisfaction, could have such a significant effect on costs that changes in the organization of labour become inevitable. (ibid. pp. 2f)

In Sweden, a committee of the Swedish employers' organization, which deals with innovation in production, has been running the "new factories" project for a number of years. During this time, new production concepts have been drawn up and experiences in setting up and running pilot schemes have been exchanged (Agurèn and Edgren 1983):

This initiative arose out of the tremendous variety of new ideas and principles which were put into practice in many areas of Swedish industry during the first half of the 1970s. The managers responsible freed themselves more and more from convention, standard textbook learning and traditional "management courses". (ibid. p. 15)

Criticism of conventional production methods is just as free from ambiguity:

Factories in which it is difficult to motivate the workers in their activities, and for which it is becoming more and more difficult to recruit, become simply inefficient in the long term both in terms of performance on the labour market and their production record. Workshops organized on a functional basis and overstocked with expensive materials, whose production planning works only on paper, deserve no respect, even if the utilization of machines under this system is slightly higher. (ibid. p. 13)

It was in England some years ago that Williamson, the father of the first flexible manufacturing system (the Molins System 24), called to account the technocentric production process at a place of prominence, namely, the Royal Society under the fitting title of "The anachronistic factory" (Williamson 1972). In this work, he establishes that "the present trends of professional management in industry and commerce [are] misconceived and [are] steadily veering into opposition with the values, aspirations and expectations of the people who will have to keep the industrial system working" (ibid. p. 140). He summarizes his criticism of workshop production as follows:

The primary aim of this method of organization, apart from the division of skills, was intended to be high machine tool utilization. The huge proportion of idle component time, and the long delay in completion, were accepted as necessary evils. As the size of workshops increased, and as the supply of skilled labour began to match the demand, sheer size increasingly defeated whatever rationale there might have been in the original arrangement. The high cost of the work-in-progress in today's conditions is sufficient reason to question the validity of these methods, but the most serious defects arise because of the lack of consideration given to two fundamental factors, communication and motivation. With increase in variety, [size] now necessitates an army of progress chasers to keep track of what is happening over the long manufacturing cycle. Attempts in recent years to replace these progress chasers and to superimpose control by computer have been unsuccessful, because of the failure to realize that the real situation is dominated by decisions, changes, omissions and errors at shop-floor level, rather than broad advance planning, and these cannot be taken into account accurately with present techniques. (ibid. pp. 142f)

In order to pursue further, the technological aspects of production, it has to be made clear that there is widespread doubt about the effectiveness of models. Model design, both in respect of the production process as a whole and for its separate procedural sequences, will not ever really succeed to the extent necessary for relatively long automated operation:

In many cases, centralized data-processing production equipment, deep inside the workshop structure, founders due to the sheer incalculable complexity of its task. Its introduction is very expensive and time consuming, and the outcome remains far short of expectations. For example, although it can register short-term disruptions of production by means of complex on-line systems, the obvious conclusions cannot immediately be drawn. (Ahlmann 1980, p. 642)

Even with the automatic determination of cutting parameters (speeds and feeds) by computer controlled equipment, only values below the maximum recorded ones can be used, in order that safety margins are not exceeded. This necessity is due partly to lack of comprehensive metal cutting knowledge, partly to varying material properties in workpieces and tools and partly to machine conditions that are not subject to calculation. Experiences in turning for example, can be cited as proof of this. They show that NC programs prepared by computer assistance in the technical office can have their machining time improved on average by 30% by skilled workers on site (personal communication; see also Kern and Schumann 1984, p. 162).

With regard to the organization of production, the objections are in the main, levelled against the rigidity of the procedures and the excessive use of co-ordination in processes where there is division of labour and centralized production planning. It applies in particular, to a great many small parts requiring numerous operations of short machining time and which make up by far the largest proportion of a parts range. These entail a disproportionately intensive use of work scheduling and production control which delays the completion of sub-assemblies and quite frequently, results in failure to deliver these sub-assemblies on time (Gauderon 1983).

Just like "over mechanization", "over structuring" can create its own problems. For example, there could be a change in the co-ordination requirements, factors that can't be quantified, losses due to disruption and the incapacity of the system to cope with improvized solutions. The hope of determining any operational activity conclusively, is obstructed by this highly complex phenomenon "operation". There are aspects of it that cannot entirely be weighed up, and being impervious to comprehensive analysis, it continually slips out of the hands of centralized accessing. (Kern and Schumann 1984, p. 163)

If the system is to work properly, there must be considerable use of co-ordination procedures. In work preparation for example, initial decision-making and initial planning work has to be carried out. Support functions (such as tools and fixtures) have to be allocated to the individual departments and production tasks have to be separated out according to the demands of the process. As a result of this, and the inevitable heavy losses due to disruption, overheads are high. The complicated flow of materials and the changeover in areas of responsibility, lead to long throughput times and the consequent need for more capital equipment. It is difficult to alter deadlines or to make changes in the details of individual items. (Ahlmann 1980, pp. 642f)

The immobility of the equipment is accompanied by the inflexibility of the staff who are employed for a narrow range of activities (Ahlmann 1980; Kern and Schumann 1984).

What comes in for particular criticism in respect of labour structures, is the inadequate use of the qualified staff that are already available, and which stems from a mistrust of skilled workers. This increases costs in two ways at once: through unexploited capabilities and through additional expense in planning and control. Moreover, this is one of the principal causes of the imbalance in manufacturing, between indirect workers and direct workers. In German mechanical engineering, the ratio of 144 : 100 appears particularly lamentable when compared with 90 : 100 which is that of our Japanese competitors who are also highly mechanized.

The training of a skilled worker costs the firm between DM 40000 and DM 50000. In addition to manual skills, it imparts the abilities necessary for planning and controlling one's own work. In the factory however, those who have been trained in this way are mostly kept on piecework. They carry out precision work on a partial process which has been stipulated in advance. This restriction of function must have a demoralizing effect. (Moll 1983)

According to the authors it leads to a "piecework bureaucracy" which, despite meticulous preparation, constant checking and re-calculation, cannot prevent:

the accumulation, after a while, of considerable amounts of spare time that cannot be made use of when piecework specifications are given (these amount to between 20% and 40% in most firms). A development of this kind can only be termed "bad news" since it brings about high costs, doubles the work, restricts work sequences and leaves untapped the reserves resulting from rationalization. (Moll 1983; see also Kern and Schumann 1984)

In view of the objective contradictions in the technocentric production concept, these arguments obviously hit the mark, but in a struggle between these two standpoints, the critics of technocentric production are not content with leaving it at that. On the contrary, wherever areas of compromise can be found in the factory, they attempt, as far as is possible, to push through alternative production structures that correspond to their objectives. Examples which we encountered earlier are the concepts of "batch control" and "workshop programming" where essential areas of authority in planning and decision-making are either never removed from the workshop in the first place or are relocated there.

In the meantime, CNC computer performance has reached a level (according to the type and extent of the equipment used) that is suitable for interaction with people, which makes programming on the spot at "the most expensive programming site in the world" seem superior, even according to the simplest production cost calculations. For example, a comparison of the costs of different programming methods used in turning has revealed that programming in the workshop, on a machine tool with convenient CNC is generally more economical than computer-based programming in the work preparation area. This applies even when the machine is not actually operating, providing there is the facility for dealing with a particular program off-line, in parallel with the machining of another workpiece (Lay and Lemmermeier 1984).

It is extremely interesting to note that comparative ease of programming is, with the market potential in mind, a major feature of the latest generation of control systems in all the important European manufacturers. Even so, most users are not (yet?) utilizing the restructuring potential of this, or the opportunities it presents for designing jobs which are suitable for skilled workers. The more clever users see the machine operator of a machining centre or production cell as the next target. They will use the human labour from mechanical processes as a starting point, since there are limits to the level of automation possible in tool and workpiece changing and in overall supervision. They intend to reintegrate as many as possible of the tasks necessary for smooth and efficient operation, by utilizing the capabilities and knowledge gained by the worker through experience. The field of activity of the machine operator should, and can, as actual case histories have shown, include programming (or at least program modifications), work preparation, tool setting, clamping the workpieces if appropriate, quality control and preventive maintenance of the machine (Kern and Schumann 1984).

However, since important problems associated with the technocentric concept still cannot be overcome by these means, more decisive and far-reaching steps are possible and necessary to bring about structural changes.

The weaknesses of conventional concepts, which include inflexibility, lack of transparency, lack of motivation and enormous expense, make it necessary to consider alternative concepts. These could be based on the following aims:

1. Dismantling complex operational, independent production units, and in particular, withdrawing from use any centralized planning and control systems, dividing up the production capacity according to the product concerned, integrating the production support jobs into these independent production units whilst exploiting the resulting possibilities of short feedback loops to correct faults, increasing the concentration of areas of authority and responsibility on site, simplifying the flow of information and dismantling the rigid and complex hierarchical structures.

2. Promoting initiatives and using human capabilities in the workshop, particularly by transferring organizational reponsibility to the individual worker dealing with the job. When there is an improvement in the flexibility of the workforce, it leads to broader and more varied work for the individual, and makes better use of his knowledge, experience and ability to react on the spot.

3. Using the most flexible means of operation possible within an organization which is also flexible. (Ahlmann 1980, pp. 634 and 645; see also Moll 1979, Spinas and Kuhn 1980 and AWF 1984)

The above implies a fundamentally different organization of the production process, a different division of functions between human being and machine and a far-reaching reintegration of tasks that previously were separated. In short, it is a way of looking at things which recognizes anew, the productive capabilities lying dormant in human beings. The production island can justifiably be seen as the nucleus of anthropocentric production. The principles underlying its organization, do not eliminate the capitalist contradiction between the struggle for control and the best production procedures, but they take the edge off it by harnessing the knowledge and capabilities acquired through experience, rather than suppressing them. Workers in the production island are not spurred on to improve their individual performance, but provided with the motivation to perform well, and at the right time. Transparancy and control, in the form of a specific scheduling framework and continual work assessments, is not in contradiction with the principles of the production island. It still remains for us to investigate the potential for development contained within it.

4.1.3 Trade Union Interests

In the past, many factors have contributed to the fact that the development of manufacturing technology and labour organization has largely served the interests of capital. So it was, that the first great wave of rationalization at the end of the 1960s, which was initiated through the use of computers and NC technology, took place during an upturn in the trade cycle. The saving in the number of workers was hardly noticeable, because job losses were balanced with job gains elsewhere. The deterioration in working conditions was hardly noticed either, owing to the many opportunities for changing jobs, or being offered "sweeteners" in the form of wage incentives. Massive propaganda compaigns, which for years characterized data processing as a vehicle for improved conditions, helped to conceal the true nature of the computer as an instrument of rationalization, and the centralization of authority. After a while, productivity increases did result in wage rises in real terms. These circumstances led to the attitude, prevalent among employers and trade unions at the time, that rationalization and technological change were basically, not to be questioned at all.

In the meantime however, there has been a clear change in the awareness and strategy of the unions, more as a reaction to the acute problems in the workplace and the consequences of massive rationalization in recent years. Of particular interest here is the attitude of IG Metall, the trade union responsible for wage negotiations in the metalworking industry. Their executive member Janzen, who is responsible for technology policy, explains:

IG Metall strongly objects to a technological "progress" which does not take into account the essential needs and demands of the employees. For us, technology exists to relieve the grind in human life. We must not allow technological progress to become a social step backwards. Two approaches to this issue should be emphasized. One is based on the need for increased protection against the negative effects of restructuring, and the other is based on the need to influence actively, both the design of future technology and company organization, in the interests of the workers. Until recently, the emphasis was on protection, and this will remain an important pillar of future policy. However, we shall have to exert a greater influence on the design of new technologies. (Janzen 1980, p. 2)

There are a number of important elements in this shift of emphasis over the regulating working methods, as well as wages and performance. These are the many agreements dealing with safeguards against rationalization which have already been signed. Two such agreements illustrating this change of emphasis are:

1. The general agreement on minimum pay scales in the North Württemberg and North Baden metalworking industry (1973) and a further agreement in Baden-Württemberg in 1978, which provide safeguards against the negative consequences of the introduction of new technologies in this industry.

2. The decisions of the IG Metall annual conference on rationalization and the humanization of working life (1980), which call for technological and organizational changes to be linked to social criteria.

The safeguards against rationalization developed so far, have to be seen in a rather critical light of course. Being essentially a social cushioning against the effects of wage cuts and job losses due to rationalization itself, they result in increased company costs. As a consequence, they contribute to the build-up of even sharper pressure for rationalization. The workers thus get caught in a vicious circle from which they cannot ultimately escape, and from whose negative consequences they are only partially and temporarily protected. Moreover, important conditions such as workloads, work content and intensity are not dealt with at all. These safeguards must therefore be urgently supplemented by regulations included in the pay agreement and by legislation, so that those involved and their representatives, have the power to influence technological and organizational changes right from the outset. An extension of workers' rights to include joint decision-making in the planning and use of new technologies is therefore the need of the moment.

But the increased pace with which many firms now pursue technocentric development, have shaken this attitude of "yes, but . . ." and transformed it into a position of "no, unless . . .". This stems from its overwhelmingly negative consequences which are now becoming blatantly obvious, such as the deskilling of jobs and job losses.

For us as trade unions, this "no" means, that we have to include in our plans, the operational preconditions under which new technologies are acceptable. This has to be considered at the level of both the individual enterprise and the national economy as a whole. In the future, the important thing will be to steer technological innovation and its industrial and social application, in the right direction. It is ultimately therefore, a question of transforming technological progress into social progress. Numerous social studies research results confirm that the shaping of work itself is more and more dependent upon labour organization and other general conditions within the company, than on the technology in use. It is therefore no longer acceptable for human capabilities to be forced to adapt to technology. We take the view that multifunctional activities, involving planning, design, control and implementation, are possible when the desire exists to make use of these capabilities and create working conditions fit for human beings. (Janzen 1983, p. 139)

The most recent expression of this view is IG Metall's campaign programme "Work and Technology – people must stay" (IG Metall 1984b). Among other things, the main emphasis of this campaign is on the design of work and technology, including a policy based on overall capabilities and qualifications, which will assist all workers.

The anthropocentric production concept involves making use of the skills and abilities of the workforce in achieving increased output and quality and reducing costs. This is logically consistent with the fact that productivity goes up considerably and leaves its mark in the form of redundancies. In addition, the extent and variety of the fields of activity offer a new chance to intensify performance. Measures for designing work in a manner fit for human beings, would therefore, if taken alone, leave employers with the opportunity of jumping in, and using the way the labour force is segmented to gain a new independence from human labour (the dangers of segmentation will be discussed later). Therefore, in negotiations, one of the central tasks of the policy must be to share out the savings that have come about as a result of this change by shortening the working week, and thus controlling the intensity of work.

Although the change in the attitudes and strategies of the unions is expressed so logically in their campaigns and documents, the industrial action is not keeping up with technological development. In practice, it is not pressure for more autonomy and for labour structures with more room to manoeuvre, that dominates the scene in union business. It is instead, a tough battle to maintain the status quo. In view of the historical development of production, this is not surprising, for just as management gets caught up in the contradictions of capital development, so the workers get caught up in the contradictions of their own position.

The capitalist production process has adapted human labour to its own internal method of operation as a result of decades (even generations) of subjugation, to the Taylorist organization of the means of production. Daily working conditions on the basis of mechanization and the division of labour gradually restricted autonomy, and no longer required skill and experience; so much so, that they finally withered away, and regardless of the limitations on these tendencies which we have already discussed, they do leave their imprint on the workers involved. In this way, the ideology of the culprits penetrated the minds of the victims in the shape of technological determinism. It imposed itself on the consciousness of the workers in the paralysing form of a dogma, that of thinking in machine terms. The production process appears to them in the guise of a technological imperative, a fact of life and the result of economic demands, but its potential flexibility and the extent to which it is all socially determined remains concealed. In this way, they internalize what capital has made of them, and as victims, they become culprits themselves by helping to create the production structure that threatens them. Instead of taking up the cudgels and appropriating all productive forces, so becoming the subjects of history as was foreseen in Marx's utopia, the workers have become shackled by the logic of capitalist development. They who through their own work, collectively develop and apply all existing productive forces, are not able to perceive them as being their own products (Gorz 1980).

As long as the workers and their representatives, fail to free themselves from the constraints of thinking in machine terms, their demand for more joint decision-making will turn into a trap in which the trapper gets caught, not the prey. Using the notion of technological imperatives, it is simpler for companies to take part in the decisions of the works council, rather than for the works council to take part in the decisions of the company, so joint decision-making would waste the opportunities those very structural changes were intended to open up. In any case, we know that in practice, skilled workers do not (yet?) lay claim to a share in decision-making on the issue of how restructuring ought to proceed. They don't entertain the idea of designing the production process in a different way from management; nor indeed, do they expect the works committee or IG Metall to be able to influence production technology and labour organization by building on alternative concepts, and making concrete design proposals.

The skilled production worker looks upon rationalization as necessary company policy, for which there is no alternative even when this has negative effects on parts of the workforce. Despite the fact that these workers are manufacturing the technology for rationalization, and that this is in part responsible for renewed mass unemployment, robbing the whole production process of its innocence, it does not result in self-accusation and feelings of reproach on the part of the workers remaining. Neither do they combine this manufacturing with demands for a different policy. Both company economics and the national economic system compel people to play along with this rationalization monopoly. In companies making machine tools, it is necessary to manufacture and sell the most up-to-date products with the most effective production methods available. The consequence of this isn't much different for skilled production workers as it is for management. Skilled workers in no way underestimate their own contribution to the production of high-quality goods, and this is what makes them accept the company's claim that producing these goods, places equal demands on everyone, and that in battling against the storm, "we are all in the same boat". (Kern and Schumann 1984, p. 215)

This attitude is fatal in the sense that production equipment and labour structures present an even greater impediment to subsequent changes and adaptations when they are closely tied up to complete systems. The more comprehensively the system is structured and the more all embracing the initial design, the fewer the occasions for direct intervention in the production system itself and the more difficult change becomes. Consequently, it is all the more essential to introduce one's own design criteria as early as possible at the conceptual stage (Benz-Overhage et al. 1982).

The displacement of skilled workers from production, even if only on a partial basis, is bound to weaken IG Metall's bargaining power. After all, they are by tradition, the core of the union's membership. However, the same process that tends to weaken it in one respect provides it with new strength in another: its support within the ranks of white-collar workers. More and more technicians and engineers are beginning to realize that the deterioration in their working conditions which leaves them isolated, alienated and treated like systems appendages, is gradually putting them in a similar situation to that of the craftsmen and the skilled workers before them. Objectively, the prospects are the same for both manual and intellectual workers. What is at issue, is the subjective readiness to work together and combine that which was previously divided in capitalist production. There are study groups of engineers already in existence, dealing with the issue of rationalization. These groups are currently instrumental in increasing the available fund of expert knowledge which can then be used to redesign the structure of production itself. An indispensable precondition for this to happen however, is that blinkered, machine-like ways of thinking are cleared from people's minds.

It is clear that one cannot redesign production along anthropocentric lines without getting involved in the turmoil of conflicting interests. The more that IG Metall includes white-collar technical workers in its membership, the more deeply will both the union and the works councils (who are responsible to the whole workforce), get caught in a dilemma over representation. If the reintegration of planning and implementation on the shop floor were to involve a significant reduction in the number of white-collar workers from the production process, there would clearly be considerable savings in those areas where production is initiated. The idea of reducing costs by getting rid of superfluous labour is of course, one of the main concerns of those in management who advocate this new production concept.

The resulting structures will be just as complex as before, but nothing will be the same as it was, even with jobs and functions still defined in the traditional way. Whatever authority the shop floor gains in respect of planning, those in the planning offices will inevitably lose. It won't be easy to escape from this dilemma. In order that life for one does not have to mean death for the other, a clearer awareness of the alternative courses of development must be a priority. What is needed is a negotiated strategy for redesign which takes the present position into account, and allows for collective learning by testing out solutions in practice.

Since the anthropocentric production concept allows for progress to be made in stages, this last requirement is indeed possible. We will return later to the prospects for dealing with this structural change.

4.1.4 Skill as a Resource

In Germany, skilled metalworkers have traditionally played an important role in the production of capital goods; more so than in other industrially developed countries, notably the USA and France (the comparison with England is less clear). Up to now, they have been able to maintain their share of production work, if not extend it. The meteoric rise in the world market of the German capital-goods industry during the last quarter of the nineteenth century was assisted not least by the productive impetus of skilled workers available in abundance, and who were trained in a variety of technological disciplines.

The position in the USA was quite different and it is no accident that it was the birthplace of Taylorism. The industrial upturn in the USA had to be exploited using large numbers of workers who were, for the most part, recruited from among the masses of black Americans recently released from the yoke of slavery into the misery of agriculture, or from among the several waves of European immigrants. The immigrants very rarely possessed any skills, or metalworking experience, and they showed little enthusiasm for working under foreign command. It is therefore not at all surprising that the enormous problem of teaching a large number of self-willed, unqualified workers who had little potential for becoming skilled and qualified, was instrumental in bringing Taylor's ideas to life, and helped his "principles of scientific management" to spread rapidly. In Germany, it was a long time before these ideas fell on such fertile ground.

Thanks to these particular historical circumstances, the Federal Republic now has at its disposal a comparatively high level of skill and capability, indicative of an unparalleled source of productivity. It is true that employment is declining in mechanical engineering as a whole, as a result of the stagnating world market and effective rationalization measures. In addition, a clear structural shift is evident in the proportion of white-collar to blue-collar workers. Nevertheless, within the dwindling group of blue-collar workers, the proportion of qualified,

Table 3. Employees in mechanical engineering, 1960–1982 (monthly average)

Monthly average or situation at end of the quarter	Total employees	Proprietors, white-collar workers and business trainees		Blue-collar workers and industrial apprentices	
		Absolute	%	Absolute	%
1960	919 170	226 863	24.7	692 307	75.3
1970	1 199 567	374 036	31.2	825 531	68.8
1974[a]	1 164 620	408 902	35.1	755 718	64.9
1978	1 067 195	382 004	35.8	685 190	64.2
1980	1 090 278	393 454	36.1	696 824	63.9
1982[b]	1 070 029	399 361	37.3	670 668	62.7

[a] From 1974 yearly average.
[b] First quarter of 1982.

Sources: Institute for Social Research, Frankfurt, "Conditions and possibilities for designing work in a manner suitable for people in the area of computer-based production processes" (unpublished interim report, July 1978); VDMA, *Statistical Handbook for Mechanical Engineering* 1979, 1982.

Table 4. Grouping structure of blue-collar workers according to sex and performance groups[a] in industry as a whole and in mechanical engineering

	Industry as a whole		Mechanical engineering	
	1970	1977	1970	1977
Men	75.4	77.1	90.0	92.4
of whom PG 1	50.2	54.5	58.8	65.4
PG 2	36.8	34.9	31.1	27.3
PG 3	13.1	10.8	10.0	7.2
Women	24.6	22.9	10.0	7.7
of whom PG 1	5.9	5.5	2.5	2.7
PG 2	45.6	46.2	42.4	38.3
PG 3	48.5	48.4	55.1	59.1
Total	100	100	100	100
of whom PG 1	39.3	43.2	53.2	60.6
PG 2	38.9	37.5	32.2	28.2
PG 3	21.8	19.4	14.6	11.2

[a] See note below Table 5.

Table 5. Grouping structure of white-collar workers according to sex and performance groups[a] in industry as a whole and in mechanical engineering

	Industry as a whole		Mechanical engineering	
	1970	1977	1970	1977
Men	65.7	72.2	68.8	74.8
of whom PG II	31.0	39.2	33.3	44.3
PG III	51.2	49.5	49.1	45.5
PG IV	16.5	10.7	16.3	9.9
PG V	1.4	0.7	1.3	0.4
Women	34.3	27.8	31.2	25.2
of whom PG II	4.3	6.6	3.5	6.8
PG III	33.2	40.4	28.9	33.9
PG IV	51.2	46.6	54.4	53.0
PG V	11.4	6.4	13.1	6.3
Total	100	100	100	100
of whom PG II	21.8	30.1	24.0	34.8
PG III	45.0	47.0	42.8	42.6
PG IV	28.4	20.7	28.2	20.7
PG V	4.8	2.3	5.0	1.9

[a] The performance groups (PGs) are statistical summaries of wage groups according to the pay agreements valid at the time. The PGs 1–3 for blue-collar workers correspond to the wage groups for skilled (1), semi-skilled (2) and unskilled (3) work. The PGs II–V for white-collar workers apply to salary groups for work involving special experience, independent performance and responsible activities with restricted authority for arrangement (II); for work involving many years' professional experience, special knowledge and abilities or special activities, and independent implementation of work but not responsibility for others (III); for work involving simple activities without the authority to make decisions, but with completed training or several years' professional activity (IV); and for work involving simple, schematic or mechanical activities without professional training (V). The PG I for white-collar activities in management is not shown.

Source: Institute for Social Research, Frankfurt, Union research project group, "General conditions of negotiations policy, vol. 2: structural data of the metal-processing, chemical and printing industries", Frankfurt/New York, 1979, pp. 233–247.

skilled labour has risen considerably. In mechanical engineering, it went from 53.2% in 1970 to 60.6% in 1977 (see Table 4). In machine-tool manufacture, its largest and most important specialized branch, the proportion rose from 67% in 1970, through 75% in 1978, to 79% in 1982. Mechanical engineering thus has at its disposal skilled, qualified labour of a clearly higher potential than that in industry as a whole, and which is increasing in relative terms.

However, there is a highly contradictory staff policy on the part of the companies involved. In general, they maintain their workforce of skilled regulars on a long term basis, and even allow it to grow, but in actual fact, they make little use of their skills. The cost of all this is made clear by the following estimates. With training costs of DM 50000 per skilled worker, mechanical engineering firms, employing around 420000 skilled workers, have invested DM 21000 million in human capital, of which it may be assumed, they will use only one third of its potential. Instead, they employ an army of indirect producers doing the planning and control that precedes and follows the production process. The ratio of workers in indirect production to those in direct production is 144 : 100. The former have essentially the same qualifications as the latter. If management were to make full use of the qualifications and capabilities of their investment in human capital on the shop floor by recombining the planning and "doing" functions and distributing them in an appropriate manner between people and machines, high savings could be achieved on indirect production work. The possible scale of these savings is indicated by the much quoted position in Japan where there are only 90 indirect workers for every 100 direct workers.

As a country short of raw materials, it certainly befits the Federal Republic of Germany to make better use of its most important indigenous resource, the abundant supply of qualified skilled labour. However, to do this in a manner which is in accordance with human needs and aspirations, it is necessary to gain an appropriate understanding of the many facets of human behaviour.

4.2 Human or Machine: Who Controls?

In complete contrast to the ideas of those who first thought of the "workerless factory" and who based their assessment of people on the performance of machines or, to be more precise, programmable digital computers, the alternative concept recognizes the fundamental differences in the basic characteristics of the human being and the machine. We have already come across some of these fundamental differences in discussing the dubious attempt to instil "artificial intelligence" into machines. If anthropocentric production is to have a chance of succeeding, it must devise a perception of human beings that does justice to these differences, and make it the basis of a design for production technology and for the organization of the production process.

As we have seen, machines are objectivizations of fixed sequential schemas of human actions, the so-called automatisms (Volpert 1984b), or, considered as models of human action, they are theory put into practice (Bednarz et al. 1984). Putting it in another way, they are the product of the human mind which is capable of conceptualizing, and the result of human action in constantly developing situations. For all these reasons, they also embody the circumstances, in particular the social ones, in which they were designed and built. Now, as soon as this design is completed, the machines, if they are to be put to effective use, place their own demands on human action. By putting their stamp on future action in this way, machines do influence the course of human lives. Production knowledge, made concrete in the form of machines, becomes a durable pattern to be passed on to future users. But, changing social relationships and situations, mean that actions have to be changed as well, and doing this leads to different machines. Earlier machines then become obsolete.

The combination of people and machines can, depending on the circumstances, turn out to have very contradictory results. If the machine dominates the operational activity, and if the human being has to subject himself as a result to this operational method, then in the long run he "withers away" to become the alienated machine operator. Although this machine confronts him as an "automatism" in concrete form, the boundary between human and machine is not external to him but internal, as a psychic entity, and the automatism generates machine-like behaviour in the human being as well (Bammé et al. 1983; Volpert 1984b). If, on the other hand, the circumstances are such that the human being is able to retain or regain his independence in dealing with machines, then he will use the machine to suit his purposes, by redesigning the objects he is working with. The machine is then a means for modifying the world around him in the most productive way. Taken together, all this clarifies the dialectical nature of the human/machine interaction.

A prerequisite for dealing with machines in a manner suitable for human beings, is a correct understanding of essential human characteristics and abilities. It is these which determine the division of functions and the design of the human/machine interaction, as well as the method of operating the machine. This understanding has to emerge out of the experience of real life circumstances.

As an individual, every human being is the product of three interrelated evolutionary processes. These are the biological evolution of mankind as a species, the historical development of his social circumstances and his individual life history. In all three processes, he lives in co-evolution with his environment which he shapes by his actions and which, for its part has a formative influence on his perception and behaviour. As a self-aware, autonomous subject, driven on by his needs and motives, he thus acts in a context which has developed with him. This context is present with him as experience, and forms the background to his current situation. From where he stands, he perceives changes in this context in respect of experiences that earlier became part of his schemas, so the schemas change as complete wholes, not in elements that fit together on the basis of rules. Along with this sensory perception and the assessment of a complete situation against the background of changing experiences, there is a continual emotional evaluation, and this is the basis for his ability to act intuitively – "because of a feeling". In addition, as a member of society, he is all the time thinking and acting with reference to his social essence and in consideration of other individuals whose social relationships, in turn, shape his experience. Communication is thus one of his basic needs (Neisser 1979; Volpert 1984b).

The schemas of his past experiences, which likewise have to be thought of as complete entities and not as a system of elements and operational rules, are not at all fully or rigidly fixed in the manner of algorithms, but are partially undetermined, so they can be altered and transferred to other situations. This enables a human being to recognize the general from the particular, and to form abstractions from the current situation, which are an essential source of his creativity. As a result, he also knows how to act purposefully even when uncertain of the outcome. The schemas are open to conceptual analysis to a limited extent. In his actions and thought processes (test activities simulated in the mind), a human being is able to remove recurrent objects and sequences of events from the particular situations in which they occur, and comprehend them as abstract concepts (objects and their relationships), so giving them an objective form (Dietzgen 1930; Volpert 1984b).

The purpose of this analysis is to form by abstraction, an analytical model of the general, recurrent elements of a class of actions that are specific to a situation. In short, it is to describe them theoretically in the form of objects and rules and then to modify them. But this depends on the ability to recognize recurrent elements in actions specific to a situation. If these are concealed from the individual human being, he can still have an internal representation of the events available, and may know how to act intuitively on this basis, but he is not able to give an objective form to them. This "tacit knowledge" is neither available in the objectivized form of abstract concepts, nor can it be communicated, at least, not between individuals whose background experiences differ (Jones 1984).

The above description should make clear the essential difference between the internal representation of a human action and the theoretical model of the same action which human beings create. Whilst an internal representation as a complete form represents experiences within the context of the individual's life history, the theoretical model arises from the conceptual analysis of individual objects or actions that have been removed from this context by abstraction. Thus it represents objectivized knowledge in the form of concepts or even data structures and algorithms. For the very reason that the circumstances of the situational context are disregarded (and in any case, not all past experience is open to this kind of analysis), theoretical models can only incompletely reflect actual events, and only insufficiently guide us when taking action.

Actions as such, are effected in accordance with a hierarchically constructed schema which is run through in sequence. An example of this is shown in Fig. 13. The basic element is a cyclical unit in which partial actions take place in order to achieve the overall action goal. The purpose of partial actions is to alter reality surrounding the individual. If the comparison between the goal of the partial action and its outcome shows that this goal has been achieved, then in accordance with the schema's hierarchical and sequential organization of action, the next goal/action unit is initiated. If however, it shows it not to have been achieved, the action is halted and the schema, if required, is changed.

The strategy for action is established at the highest level, where the overall schema of the action is generated. This schema, not entirely determined in all its details, has the intention of achieving the overall action objective. The motor–sensory goal/action cycles exist at the lowest level. In between, there can be cyclical units on several levels, where partial goals are generated as well as the schemas of the partial actions, allocated to the subordinate goal/action units (Österreich 1981; Volpert 1975; Volpert 1984a).

This view of a human being, this model of his thought and action, enables the contradictory capabilities of people and machines to be clearly understood. A human being can act, even in an unfamiliar environment, but in processing

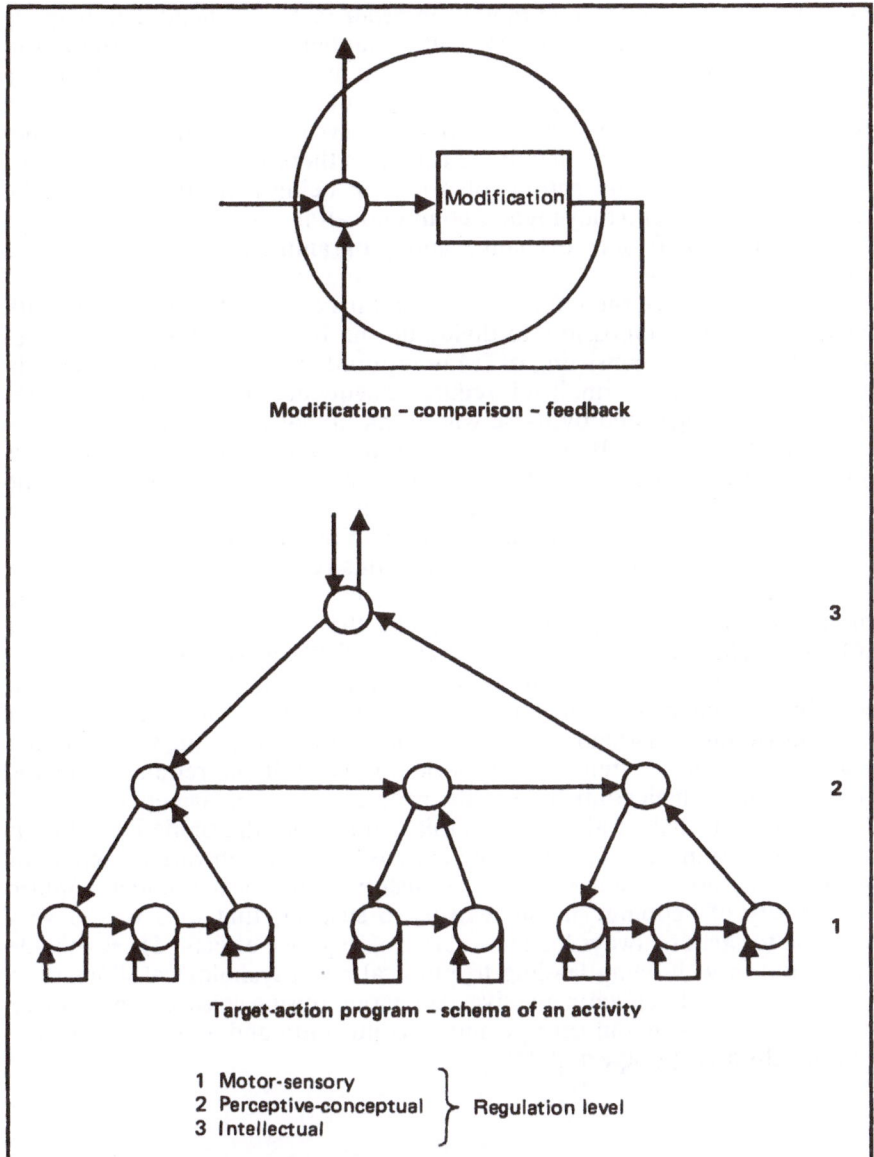

Modification – comparison – feedback

Target-action program – schema of an activity

1 Motor-sensory
2 Perceptive-conceptual   } Regulation level
3 Intellectual

Fig. 13. Action regulation (after Hacker, Volpert).

information, his capability in both speed and memory capacity is not at all impressive. A machine can offer an enormous data-processing capability, and there are possibilities for knowledge processing under development. To transform input data into output data however, programs are needed which are drawn up beforehand and can only be the product of human intellectual work. A human being is part of a complete situation and is capable of overall assessment, whereas a machine needs abstract models in order to comprehend data from its environment analytically, and process it in a manner befitting the situation in hand (as far as the models permit). The flexibility of the schema enables a human being to act purposefully, even without rules, and to be creative in an unsystematic and inconsistent way, whereas the systematic and consistent behaviour of a machine is strictly tied to the rules and therefore neither flexible nor creative. The intentions and actions of a human being actively challenge the world in which he lives, and they involve communicating with his fellows, whereas a machine is limited to data exchange and programmed interaction with its environment.

The special qualities of the human being, and his dialectical relationship with the machine, makes it impossible to divide up the functions of a work process into two separate parts, consisting of (a) unfettered creations of pure reason in the human mind, and (b) formalized, regulated sequences in the computer. This is the sort of thing suggested by those who think in machine terms, and reckon they can take a short cut. If people are to retain a way of working which is appropriate for human beings, and if their abilities are to be combined with the contrasting capabilities of the computer, then it is much more a question of designing the division of functions and the interaction between human and computer, according to criteria which accord with this relationship.

Using this view of human beings as a starting point, we can derive some general criteria for job design which will make the work involved suitable for human beings. These criteria are summarized in Table 6 (Hacker 1978).

Since the personality of a human being develops in the context and circumstances of his particular life, and since a large part of life consists necessarily of work, the individual's personality is shaped by his working life quite substantially. In particular, it is the content of the work, the capabilities required and the social relationships which contribute to the promotion or impairment of personality development (Ulich et al. 1980). In this connection, the scope for action is of particular importance. There are three dimensions to it: job variety, scope for making decisions and varied social relationships. Empirical findings confirm what the theory of "coping" might lead us to suspect, that jobs with a very narrow scope for action always lead to a series of negative effects. These include a deterioration in well-being leading to physical and psychological illnesses, a decline in intellectual capacity, passive behaviour during leisure time and an educational attitude which in turn promotes conformity and a lack of independence in the children (Volpert 1982).

Table 6. Criteria for a work form suited to human beings (from Hacker 1978)

Assessment level	Criteria (examples)
Personal development	Freedom of action (work tasks, decisions, communication with colleagues)
	Training
Demands	Stresses (physical and psychological)
Security	Stress maxima
	Employment security
Feasibility	Ergonomic norms
	Limits of perception

There are a number of design principles that should be noted when dealing with the interaction of human and computer. Existing scope for action in the work situation, can only be exhausted if the initiative comes from the person who determines the work sequence, evaluates the intermediate results in context, and makes the decisions on appropriate action. In order to avoid undue difficulties, the functional division between human and computer must enable the human being to carry out complete operations which place demands on the way he copes with it. The demands have to cover the three dimensions of his scope for action so that he is able to use the computer as a tool instead of just operating it. The functions and modes of behaviour of the computer in the production process must be completely understood by the person involved, and the response of the computer to certain input data must be clear to him in a form appropriate to the particular situation. In combining human being and computer, what is crucial to the success of the person's performance, is his ability to recognize and understand in detail, the connection between the stages of his work and his intentions, especially his own inputs and their effect, both on the workpiece and the whole work situation itself. Only in such circumstances would a human being be able to perceive in an overall context, the effects of a particular action in a specific situation, and be able to cope with the whole activity, thus retaining his integrity and acquiring new capabilities.

It is hardly necessary to point out that computer systems at present on the market do not conform to these principles. They were after all, developed for processes which are heavily dependent on the division of labour and in any case, are functionally inadequate in general. The best that can be said on that score, is that some progress has been recorded recently in designing the work in a way which is more suited to their users. In the meantime, certain design rules have been worked out in detail, on the basis of industrial psychology, for use as guidelines in the design of equipment suitable for people to use. They are not just concerned with the work performed by users or with the human/machine interaction, but also with the attendant work organization (Spinas et al. 1983).

4.3   Group Technology: A Non-Taylorist Route

4.3.1   The Principles of Group Technology

The principles of group technology are almost as old as Taylor's principles of scientific management on which the opposition is based. Following the November Revolution of 1918, the hopes and demands of German workers were largely for the production process to accord more with their interests as producers and less with the division of labour. At that time, they had already put behind them their first experiences of rationalization, which were in mass production and in war production. These were based partly on Taylor's principles (at the time they were being translated into German). Even when later, with the promise of higher wages, these same principles prevailed, and determined the course of events, many businesses were persuaded by the political balance of power and the arguments of the workers, to pour oil on troubled waters by using alternative production concepts. This was what led Lang, the production engineer at

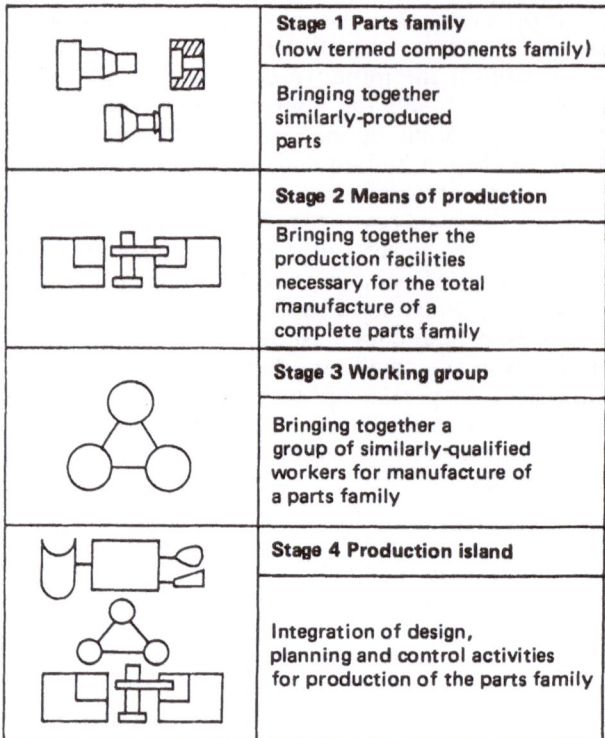

	Stage 1 Parts family (now termed components family)
	Bringing together similarly-produced parts
	Stage 2 Means of production
	Bringing together the production facilities necessary for the total manufacture of a complete parts family
	Stage 3 Working group
	Bringing together a group of similarly-qualified workers for manufacture of a parts family
	Stage 4 Production island
	Integration of design, planning and control activities for production of the parts family

Fig. 14. Principles of group technology.

Daimler–Benz, and Hellpach, the industrial psychologist at TH Karlsruhe, to work out their concept of group production, which already embodied all the principles of group technology (Lang and Hellpach 1922; Neubauer 1981).

At the core of group technology is a completely different way of organizing production following the rule of splitting orders instead of dividing labour. For instead of dividing up work into particular specialist tasks, the range of orders is grouped into families of parts which have similar demands on production. Instead of dealing with all the production orders in separate routine tasks, each family of parts is produced as a whole. Thus the old style workshop production, organized on the basis of the division of labour and routine tasks, turns into group technology, organized on the basis of similar requirements and total manufacture. In this latter method, the production process is reintegrated in two respects. Firstly, the factory orders are carried out in full as complete wholes on equipment which has been combined both spatially and organizationally, and secondly, human work is organized in teams with complete work sequences. To be more specific, group technology is founded on four intertwined principles as illustrated in Fig. 14 (Lang and Hellpach 1922; Mitrofanow 1960; Williamson 1972; Ahlmann 1980; Warnecke et al. 1980).

Management frequently doesn't go beyond the first stage and contents itself with grouping the range of parts into families. This is a customary rationalization measure for bringing more order and clarity in respect of drawings, parts lists and work schedules. Without doubt, this enables costs to be reduced by searching for similar parts, drawing up variants and cutting down on the range of parts to be manufactured. Although its prospects often go no further than this, group technology is still representative of an overall concept for organizing the whole of production. It is not just an isolated restructuring measure. As we know from experience, its most important economic effects will only be felt when there is the opportunity of putting all its principles into practice, so that its potential can be fully exploited.

Group production, which embodies all the principles of group technology, has the following advantages over workshop production (Williamson 1972; Ham 1977; Warnecke et al. 1979; Spinas and Kuhn 1980; AWF 1984):

Throughput times are reduced dramatically. The reduction is between 60% and 90% when production islands are set up. The corresponding reductions of work-in-progress is between 30% and 60%.

The expenditure on work scheduling and NC programming is lower, as a result of their transfer from the technical office to the production island.

Members of the team have a greater variety of work and more scope for action. There is the opportunity of spreading the workload and developing new capabilities.

The planning and control work is greatly simplified because each production island functions as a separate production unit.

The Achilles heel of group production is clearly its unbalanced utilization of capacity, which is in complete contrast to the high utilization in workshop production, its only economic strength. The production capacities of an island cannot just be adapted as required to all the parts families, but the economic drawbacks can be alleviated to some extent even though they still exist. Firstly, the work must be arranged to make full use of the most expensive machine (the lead machine). Then, to balance out a lack of capacity in one production island with excess capacity in another, it is feasible to make use of free capacity to a strictly limited extent. Additional capacity can be gained on machines where the work has built up, by shortening preparation times and selecting a suitable sequence for the orders.

Group production workers are more highly qualified than those employed on specialized types of work and are paid more. The additional labour costs however, are easily offset by the other economic advantages. At a comparable level of automation and computer use, the costs of software development and software maintenance are lower. This is because there is much less work to be carried out on programming model designs and adaptations, and this has a desirable effect on capital intensity. With this form of organization, overall productivity increases considerably. Obviously, an economic assessment is only possible in individual cases on the basis of a precise cost–benefit analysis. Nevertheless, even in this general examination, the features of group production's economic superiority over workshop production are clearly evident from experiences with production islands already functioning.

When the principles of group technology are applied, they lead to substantially better working conditions than the technocentric concept of production does, even when they are measured against the very best standards of job design. In production islands, a large number of very varied tasks have to be carried out in succession. They proceed from work scheduling, programming and organizing materials and orders to the use of different items of production equipment such as machine tools, operating equipment and computer systems. Assuming that all members of the team have the same basic capabilities, their co-operation is essential in uniting the planning and implementation functions. Their scope for action in all respects, is widened considerably, and offers numerous opportunities to develop new social and technological capabilities. In addition, the workloads arising in respect of individual tasks can be shared out (Spinas and Kuhn 1980).

4.3.2   Integrated Production: Component Family Production and Design Islands

In production, there are two fundamental areas of knowledge. One is the knowledge of how a machine works, and this is incorporated in the design; the other is the knowledge of how the machine is made, and this is established during the production process. It is true that the two spheres are interdependent but, again and again, powerful forces cause them to be distinguished. Probably the most basic difference is that in design, it is a model that is being handled, whereas in production it is the real thing. It would seem worthwhile redesigning both areas jointly, according to the principles of group technology, and thus doing justice to each aspect by assisting the workers in obtaining, processing and utilizing the knowledge they have gained.

To consider production and production planning on the one hand, as a mere appendage to design or as a vestige of it, as they do in technocentric production, would be inappropriate. On the other hand, the incorporation of design into production wouldn't be suitable either. The two areas have to work in close collaboration with each other, and this must be done by technological and organizational means.

If the principles of group technology are used in production, it usually leads to a production island. The Committee for Economic Production (Ausschuss für Wirtschaftliche Fertigung – AWF) defines it as follows:

The task of the production island is to manufacture either parts or end-products as completely as possible, starting with basic raw materials. The required means of production are organized and spatially arranged in the production island. The group's area of activity displays the following characteristics:

Extensive control by the group members over their own work, and co-operation between group members on planning, decision-making and control functions within a set of specified general conditions.

The renunciation of an over rigid division of labour, and a consequent widening of the scope for re-arrangement on the part of each individual. (AWF 1984)

Just as production islands are formed for the complete processing of whole components families (previously termed parts families), so design islands can be formed as well, on the same group technology principles. A prerequisite for this is that the products are well structured and built up from sub-assemblies (which is economically desirable in any case).

Instead of dividing up design along functional lines according to the design stage, which is a custom now on the increase, the products and subassemblies are put in batches to from families. Every family of products or sub-assemblies thus formed, is then subjected to the complete design process in a design island to which it is allocated. All stages of the design process are carried out in this case,

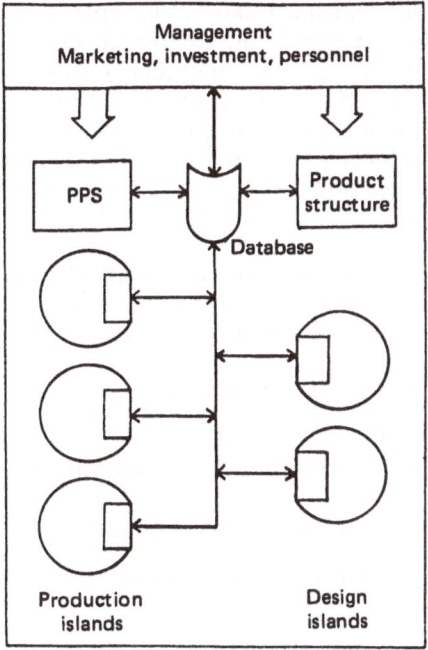

Fig. 15.   Integrated group production.

right from functional specification, through initial design, calculation and detailing, to drawings and parts lists. The group of designers who have the job of
carrying this out must have a very wide range of qualifications and abilities to
call upon, and should all be equally capable, so that design requirements can be
dealt with in their entirety. In this work, the designers have the support of local
computer systems in their area.

   In this way, two production subsystems are created which operate relatively
autonomously. They are the production island and the design island, organized
around a core of specific capability and supported by computers. According to
the anthropocentric production concept, they form the basic production units in
the factory of the future. These basic units must be able to exchange data to
work effectively, for although they have relative autonomy, there are functional
interconnections between them. For this purpose, they need the basic components of CIM design which are: a common database, a data-transfer system
and compatible data interfaces as shown in Fig. 15. If for example, an NC program
is to be drawn up in a production island and the geometrical data structures have
been prepared in the corresponding design island using a CAD sysem, then
it should be possible for these structures to be called up from the database.
Examples of this might be a contour description in turning, or drilling details for
a machining centre. It must be possible as well, for data connected with the
orders to be transferred to the production island.

   It is important to note that the computer support is quite substantially different
in essence, from computer use in the technocentric method of production. For
example, the programs have to fulfil quite different functions corresponding to

the new role of human labour in production, and must satisfy the criteria of designing work suitable for people. Instead of objectivizing almost all knowledge and work sequences as far as possible and giving them an objective form in the computer system, the design island should instead, support the designers and skilled workers in creating their own tools and means for doing the work. For in this instance, the computer system distributed locally is a general up-to-date and consistent information system with which routine operations can be carried out on the spot, although the planning and decision-making has to be left to the qualified personnel.

Since large parts of the planning and co-ordination activities are established in the production and design islands, the more general tasks of design and production planning and control, contract down into comparatively simple core systems.

The function of production planning and control is now limited to:

1. Converting customer orders into production orders
2. The costing of materials and the planning of total capacities and deadlines for the small number of islands
3. Maintenance of the CAM (computer-aided manufacturing) systems

The more general tasks of design and development essentially include:

1. The plans for general development
2. The configuration of products
3. Maintenance of the CAD (computer-aided design) systems

It is nevertheless intended that these more general functions, which have been reduced and greatly simplified, will play an important role in ensuring the efficiency of the whole production process. With their attention firmly attuned to dealing with orders according to the customers' wishes, they have to co-ordinate the work of the individual production islands in such a way that overall, they can guarantee the lowest stocks and throughput times, as well as high quality and output. The autonomy of the islands must be linked to consistency and stringent co-ordination, so that the factory which has been redesigned along group technology lines, can play its economic trump cards to the greatest effect.

4.3.3   Shop-Floor Programming: A Tool for Skilled Workers

In order to demonstrate the importance of having technical aids that are appropriately designed for human use, we shall take another short look at NC programming. We have already seen that decisions on how it should be used depend mainly on a firm's organization and staff development strategy, rather than on technical factors. Surprisingly it often appears preferable for the skilled workers on the shop floor to write the programs, rather than the technical office personnel. It usually results in lower costs, less time being needed for setting up and testing, and the production of more efficient programs. This is because the practical experience and knowledge concerning the status of machines, tools and fixtures, are available on the shop floor whenever they are required (Lay et al. 1983).

In contrast, traditional abstract programming and editing techniques never give a thought to the way a skilled worker thinks and works, and so they can hardly be said to be suitable for them to use. Skilled workers know how to read drawings and transform the information they contain into the control of machine tools. As discussed earlier, their knowledge is based essentially on practical experience, and so it is concrete, graphic and analogical in nature. Traditional programming methods however, stem from the realms of engineering and data processing, and are therefore abstract, algorithmic and digital in nature.

As a result of these findings, new shop-floor programming methods have been developed that are suitable for the processes of turning, milling, grinding and sheet-metal work. This has been done within the framework of the Federal Manufacturing Technologies Programme. Despite the inevitable process-specific features, they are provided with a unified user interface which assists the skilled workers just like a new tool. It allows them to make use of their manufacturing knowledge, their planning capabilities and their knowledge of the status of machines and tools to produce efficient, optimized NC programs. By doing this, skilled workers can reproduce their own technical competence themselves.

Accordingly, functions and tasks are allocated to both human and computer in such a way that it is up to the worker to plan the program's flow and to decide on the cutting data required. This is in complete contrast to earlier attempts to determine operation sequences and cutting parameters automatically. The worker thus maintains the initiative, evaluating working situations and deciding on the actions to be taken. The programming system is designed to assist him by keeping recurring geometrical elements and cutting cycles available in a generalized, pre-programmed form, so that he can build up the program by setting parameters and combining the modules in the right order. Further assistance is provided by a tool database and, most importantly, by graphic real-time simulation of the tool paths as a means of testing the programs.

Because skilled workers have a concrete and graphical way of thinking and acting, the design of the user interface is based on diagrammatic representations of the objects and operations needed for a program that are very similar to those used in drawings. This is in contrast to common abstract and symbolic programming languages. As an example, the interpretation of data formats is portrayed diagrammatically (Fig. 16).

Apart from this, the programming procedures meet the requirement that they must be easy to follow and self-explanatory, so that they continually indicate to the user, the current state of the system, the actions open to him and the effects of these actions. They allow him to correct a sequence of instructions that have resulted in errors.

All in all, this programming system makes the most of the performance capabilities of modern computers, especially where the user can interact with the software by direct manipulation or by means of "window" techniques. It proves to be a complex, but still suitable tool for skilled workers. Existing skills and knowledge are thus made use of and maintained, rather than being replaced through "intelligent" software (Liese 1989). Experience gained from the use of this programming method indicates, that measured in terms of programming and testing costs, it is far more economic than any other known method. Moreover its suitability as a tool for skilled workers is evident from the very favourable learning curves achieved during training. After no more than fifteen minutes instruction, the system can be used unaided, and after a five-day period of learning, even skilled workers with no previous programming experience perform as well as their experienced colleagues (Liese 1989).

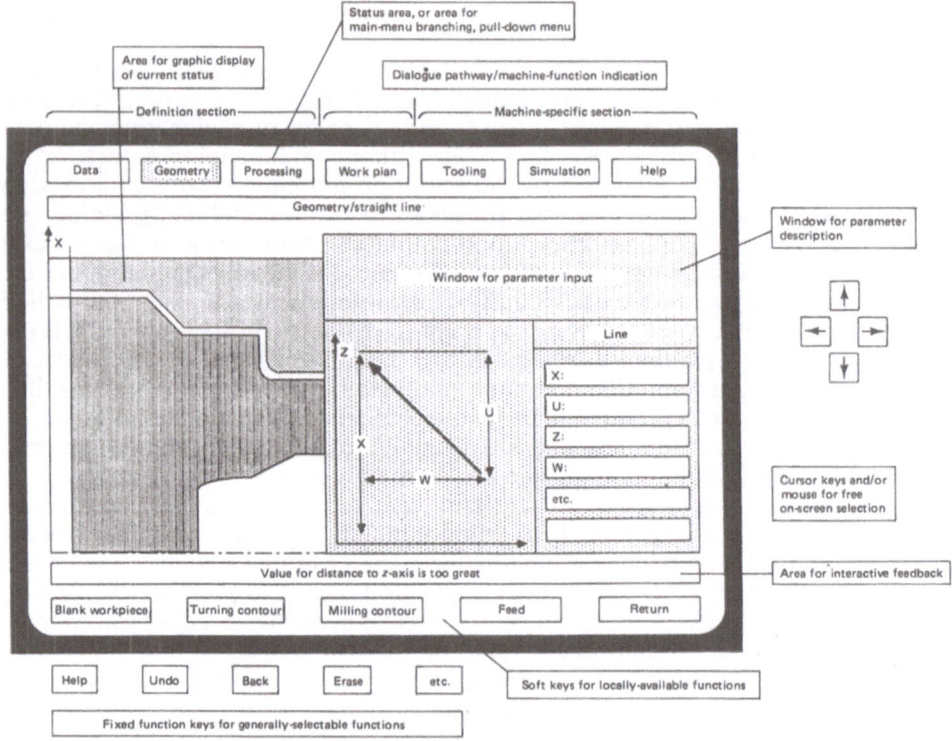

Fig. 16. Basic form of interaction.

This illustrates how important it is to chose the right perspective in systems design. Technical systems cannot be designed appropriately without taking account of the circumstances of work design and without having due regard for the relationship between work organization, skills and qualifications on the one hand and the functional requirements of the technical system on the other. Human endeavour has to be recognized as the primary source of productivity even though it causes failures and raises costs. Efficient production might require a search for new ways of combining the unique capabilities of humans with the performance of machines. Therefore, unlike traditional attitudes, the perspective in systems design must sometimes depart from computer programs which imitate and replace human expertise and envisage the design of computers as a useful tool for human experts. The "computer-aided-craftsman", not the computer, could emerge as the new hero of production (Jones 1984).

4.3.4   Different Forms of Component Family Production

At the present stage of development in production engineering, there are three types of machine systems with which components families can be manufactured:

1. Machining centres: these contain all the machining processes required for total manufacturing, all combined in one basic machine.
2. Production islands: these are formed by combining conventional individual machines both spatially and organizationally, and also combining the methods of operation necessary for total manufacturing.
3. Flexible manufacturing systems: these are a special type of production island, in which the flow of items between the machines is automated, and the whole machine system is controlled by means of a computer.

Machining centres have assumed their classical form in processing centres for drilling and milling prismatic workpieces. Over recent years, various efforts have been made to expand certain types of lathes into turning centres by using live tooling (for example, for drilling and milling operations). In the present context however, this form of production for components families has two serious handicaps. Firstly, the comparatively small tool magazines and the expensive devices for changing workpieces, restrict the operation to a small range of parts. Secondly, in order for the range of parts to be expanded, long preparation times on expensive individual machines have to be tolerated, so this is why their use remains restricted in the way described. On the other hand, these barriers can be overcome if two or more such processing centres are connected and provided

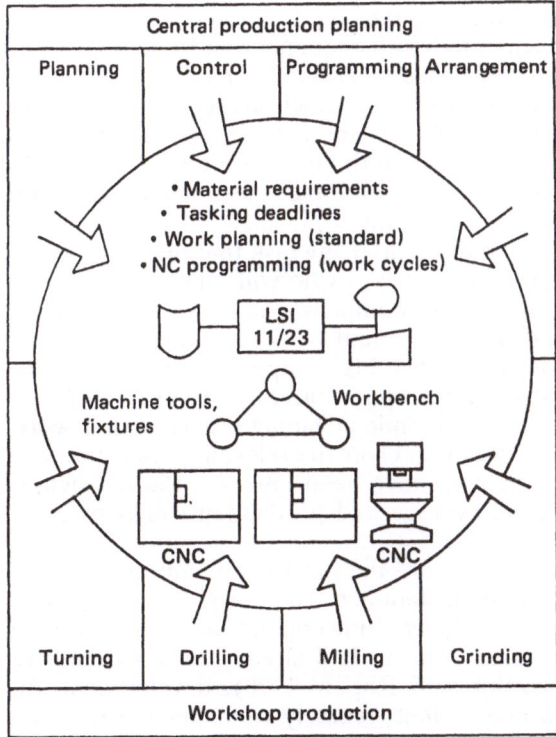

Fig. 17. Example of a production island.

with tools from a large common store (Hammer 1983b). We now have a flexible manufacturing system with machines which are largely interchangeable.

Production islands are the most general form of integrated manufacturing of components families. They can be implemented even without costly automation equipment. Reports on their economically advantageous use are frequently published, although this is still confined mainly to individual parts families (Williamson 1972; Warnecke et al. 1979; Mönig 1980; Zöller 1980; Bittcher 1981; Jakob 1981; AWF 1984). In Sweden however, there are already whole factories in which production takes place entirely in production islands (Aguren and Edgren 1983). A more detailed example developed within the framework of the production engineering programme, may clarify the details of the concept (Gauderon 1983; Autorengemeinschaft 1984).

In a medium-sized mechanical engineering firm with conventional workshop production and centralized work scheduling, a production island was set up for the complete production of a range of 4000 different parts (Fig. 17). It is comprised of a CNC lathe, a conventional lathe, a CNC drilling and milling machine and a place for working with hand tools, as well as all the tools, devices and measuring equipment required for production. A group of three equally qualified, skilled workers works in the production island. Their tasks include arranging and preparing materials, controlling the production orders in accordance with the deadline (this particular batch of orders was specified ten days in advance), preparing work schedules and NC programs, and ensuring quality. A computer (LIS 11/23) is at their disposal in the island to assist them in these tasks.

The work schedules are derived from standard ones, worked out earlier and stored in the computer and supplemented with sequences and data which are specific to the current workpiece and process. A similar method is used for programming generalized manufacturing cycles, which are completed by the input of specific geometrical and technological data such as measurements, cutting speeds, feed and depth of cut. These methods correspond closely to the normal working procedure of skilled workers and make use of their specialist knowledge, being very efficient at the same time.

Since every member of the team can undertake any task that may arise in the production island, they decide among themselves who will take on which task next. For this too, they have the support of complete and up-to-date situation summaries in the computer, and can take account of circumstances which are dependent on the situation in order to achieve high output and overall quality. These circumstances might include cost of preparation, availability of tools and fixtures, and machine conditions. The economic advantages over earlier workshop production are clear from this example. Over the relevant period the costs of work scheduling and NC programming have been lowered, the throughput times have been reduced by 70% on average and production stocks by 30%. Productivity has almost doubled.

Flexible manufacturing systems (FMSs) are the most highly automated form of integrated manufacturing of components families. Their origins go back to the 1960s when in 1967, the Molins System 24 went into operation as the first FMS (Williamson 1967). Although the basic concept had already been worked out systematically much earlier (Dolezalek and Ropohl 1970), development and expansion took a hesitant course and at first, scarcely went beyond tentative efforts and elegant systems of dubious economy. A breakthrough on a broad front was recorded at the start of the 1980s initially in Japan which in 1982 had

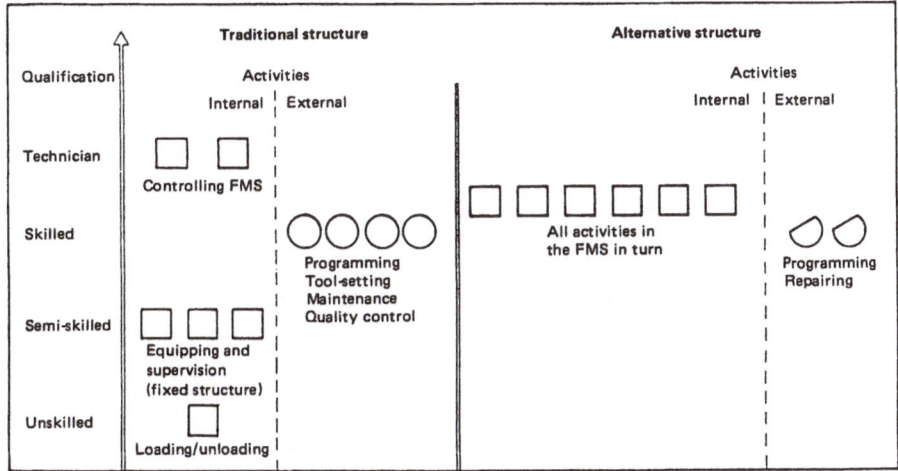

Fig. 18. Alternative work structures in flexible manufacturing systems. Source: Institute für Sozialwissenschaftliche Forschung, Munich, 1983.

thirty installations. At that time the USA had fifteen installations and West Germany had twelve. (Warnecke and Steinhilper 1983b; for further summaries see Spur and Mertin 1981, Hatvany 1983, and Schulz and Arnold 1983).

Essential parameters which determine the success or failure of an FMS, are its flexibility as measured by the variety of parts in the components family to be manufactured, its usually high use of linking facilities such as storage, transportation and handling equipment and the work structure. Most FMSs are designed for fewer than ten different workpieces. Their low flexibility requires fewer machine tools and less use of interconnections. However, there are also systems with substantially higher flexibility (over 200 different workpieces) and with more than twelve machines connected together (Schulz and Arnold 1983; Warnecke and Steinhilper 1983b). The efficiency of the FMS is entirely dependent on its overall availability and on its output, and it can be increased through the use of monitoring equipment. Above all, it is influenced by the staffing structure and by staffing policy (Lutz and Schultz-Wild 1983).

Empirical findings show that contrary to all expectations, the availability and output of the FMS are often disappointingly low. This can be attributed to a sharply polarized staffing structure such as the traditional structure illustrated in Fig. 18 (Gerwin 1982).

Nevertheless, despite few differences in staffing policy and staffing structures, the traditional polarized structure is generally found wherever the FMS is used. The exceptions are two German systems and several Japanese ones where great value is placed on the flexible use of personnel with equally high abilities (Autorengemeinschaft ISI, IAB, IWF 1982; d'Iribarne 1982). Although FMSs used in this way are based on the first two principles of group technology, they should be seen as manifestations of the technocentric course of development. With them it is the intention to go beyond the boundaries of individual automated machines and achieve the breakthrough into the ghost shift, as long as the

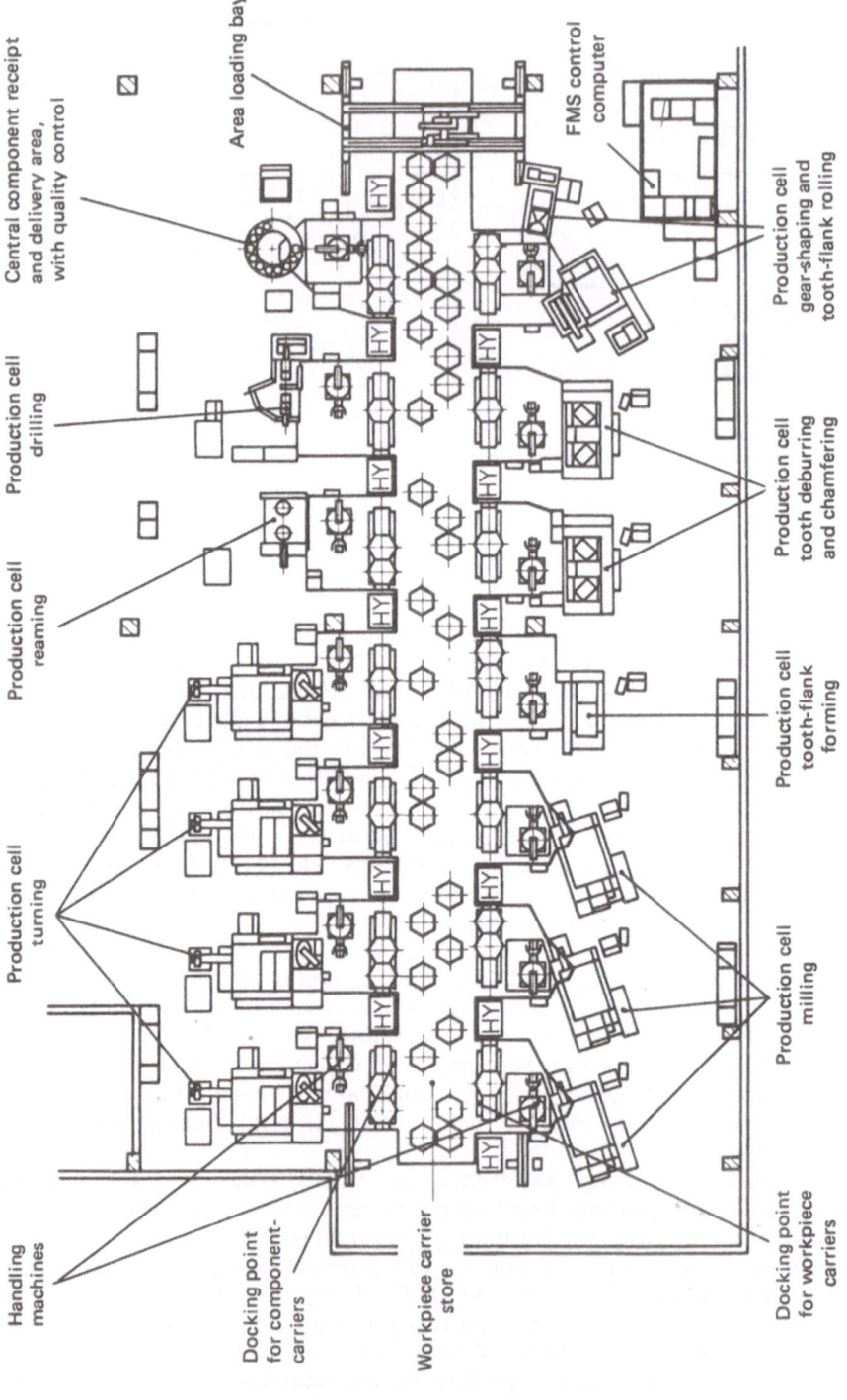

Fig. 19. Flexible manufacturing system for machining axially symmetric components of diameter 60–280 mm (courtesy of Zahnradfabrik Friedsrichshafen AG).

problems of process monitoring can be solved. They are the technocentric thorn in the flesh of the anthropocentric concept. However, problems with availability and output of FMSs could in the future, lead to the remaining principles of the anthropocentric concept being given more recognition.

In any case, it is clear at this point, that between the two conflicting and idealized courses of development, there are various middle ways. A more detailed look at one FMS, developed within the framework of Manufacturing Technologies Programme, may clarify this further (Hörl 1982; Köhler and Schultz-Wild 1983).

The flexible manufacturing system (Fig. 19) comprises all the machines, tools fixtures and programs that are necessary for the complete manufacture of a components family consisting of over 350 different, rotationally symmetrical parts (gears, sleeve bearers and coupling parts) in batches of between 50 and 500. The modular construction enables the production cells to be operated autonomously so that FMS can be dealt with in stages. Each of the 14 production cells contains a machine tool, a handling device and a positioning unit for the workpiece magazines with a common control and measuring and monitoring equipment if appropriate. The operational areas were designed to be as uniform as possible and the working and service areas were strictly separated. The workpiece magazines are kept in a central store with random access and are transported using a portable handling device. The whole system is controlled by a process computer (PDP 11/45) with programs for NC program administration (DNC), monitoring and diagnosis, control of the flow of materials and the human/computer interaction.

In contrast to other FMSs, the labour structure is such that all those working in the system are able to carry out any activity that may arise as illustrated in the alternative structure in Fig. 18. The exceptions to this are the few tasks carried out by the head of the plant. The workers receive instruction in the various production and measurement procedures as well as in the control in order to prepare them for this work. The elements to be learned permit the teaching method to accord with the initial capabilities of the workers, who could be semi-skilled or skilled metalworkers.

A variety of staffing structures were evaluated in a discussion between the management and the works council, on the basis of industrial engineering and social science analyses, and measures for increasing the capabilities of the workers were agreed. In opting for a labour structure in which all workers are able to replace or take over from each other, a decisive factor was that high availability of the FMS is only feasible with a flexible team whose members are all as highly qualified as each other. Compared with the high capital costs, the additional costs of employing more highly qualified workers look rather modest. For the workers, there was much wider scope for action, and productivity was increased by 66% over conventional production.

The general connection between investment, higher efficiency and the allowable increases in labour costs can be calculated. The amount that may be spent on personnel costs in order to increase the extent to which capital-intensive plants are available for use, can be seen from Fig. 20. For example, with investment costs of DM 4 million and an annual increase of 60 hours in operating time, an increase of DM 20 000 in personnel costs can be justified (Seliger 1983, pp. 33f).

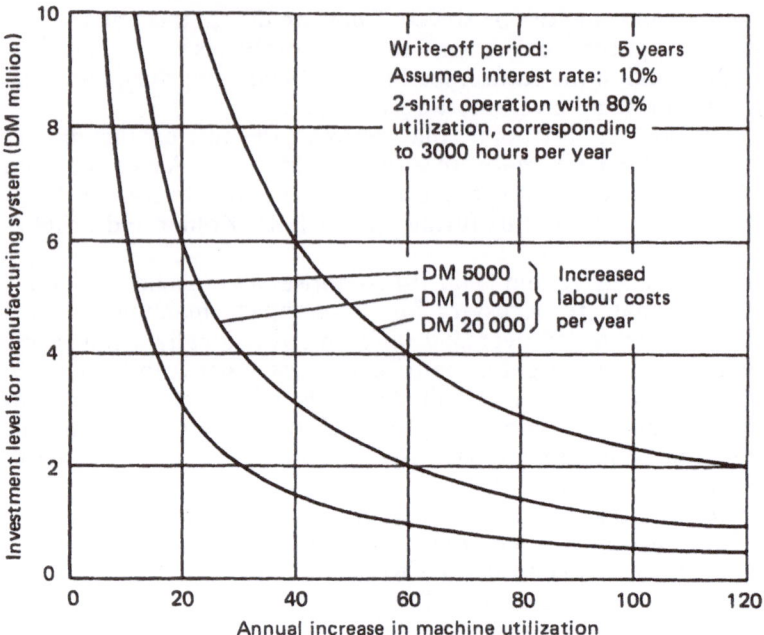

Fig. 20. Equivalence curves for additional unit labour costs and increased machine utilization.

4.3.5 Design Islands: A New Structure for Design Work

Traditionally, the structure of design is heavily dependent, for a number of reasons, on the nature of the object being designed. The whole nature of the design process itself calls for an integrated procedure. For the individual objects must perform their function, but in combination. They have to act as a complete whole in the form of a machine. But the abundant and comprehensive experiential knowledge, which ultimately determines the quality and efficiency of the product, despite all theoretical generalizations, is still only called upon in certain areas of application. It must be said though, that as products increase in complexity and design tasks grow in range, structuring principles embodying the functions involved in design are beginning to spread too. The application of these principles is bound up with the division of labour, and with the polarization of skill requirements. They have not up to now, made as much progress in Germany as they have for example, in the USA.

  In contrast to this, the CAD systems on the market are entirely concerned with the automation of individual functional areas of design as we have seen. As instruments for isolated reorganization efforts, these CAD systems presuppose a pronounced functional structure so that they can display their limited economic effects. They thus compel companies using them to reorganize in a way that is of little use to the design process itself, and fragments it unnecessarily. Although the isolated rationalization of certain functional areas does save time and money within these areas, the consequence of doing it, is to undermine economic efficiency of the overall design process itself.

It is quite true that no detailed studies on this subject are available, but there are justifiable grounds for expecting redesign along group technology lines, to bring about similar favourable effects in design as have been demonstrated in production itself. This of course, requires completely different forms of computer support from those offered by the CAD systems currently in use or under development, and they will need to be designed in an entirely different way. The RUKON system, promoted under the auspices of the Manufacturing Technologies Programme, does much more justice to the anthropocentric concept, and can demonstrate what is involved.

Instead of automating partial functions of the design process and using programmed sequences to displace the designer from it altogether, efforts are being made to give appropriate computer support to design as an integral process, and to leave the organization of the work and the control of its procedures to the designer. It follows that computer support does not begin at the stage where it is simplest to express ideas as algorithms, but during the initial design of the products and sub-assemblies. Conceived independently by the designer, this design is stored in the computer, where different versions, even of details, can be retained and recalled at will. At this initial stage, any desired calculation programs can be used from the same workstation. For example, calculations for housings or shaft couplings. In this process, the dimensions of the design which were fixed on a provisional basis, are used simply as an input and the results of the calculations are then incorporated into the design.

After an overall scale design has been drawn up in this way, the individual parts can be taken from it. They can then be completed in the form of diagrams and section drawings, reworked with the production methods in mind and converted into drawings and parts lists suitable for manufacture. Consistency is maintained throughout in fixing tolerances, because the overall design framework for the individual parts is in the computer. At the end of the design process, drawings and parts lists are drawn up automatically.

For the individual designer or group of designers, working with a system of this type means that they have once again, the experience of running through the design process as a whole, and thus bringing their own specialist knowledge into play, gathering new experiences and having a wide scope for action. In addition to this, reaction from industry on initial experiences in using the system, indicates that its efficiency is high as well. At present, the system is restricted to two-dimensional work and a great deal of further development is necessary to turn it into an instrument that can be extensively used in all areas of design. It seems worthwhile pursuing this course further, rather than getting caught up in those technological and organizational bottlenecks of rationalization from which production is only now attempting to escape.

4.4   Forces of Inertia

Although the economic advantages of anthropocentric production are obvious, powerful arguments are preventing decision-makers from pursuing this course. We shall discuss them in the order of the scale of their influence.

1. Going into group production requires a great deal of preliminary work in analysing the products and the range of parts in order to be able to structure products and form parts families.

In every case however, efforts to rationalize are needed to bring order into production, counteract the proliferation of parts, and provide an economic benefit already taken for granted. They do not therefore, present a serious obstacle.

2. On the basis of group technology's second principle, the means of production required for the total manufacture of a parts family would have to be spatially combined. This would mean rearranging machines and disbanding central installations which are part of the current means of production.

In view of the frequent rearrangements taking place in the factory anyway, and bearing in mind the fact that it can take place in stages, components family by components family, this cannot be seen as an insuperable obstacle either. Special provision will have to be made in the rare cases where it is not possible, such as at hardening plants.

Furthermore, group technology requires distributed computer hardware. Foreseeable trends in the microcomputer market's price/performance ratio, and in the development of local networks, likewise mean that this requirement presents no insurmountable obstacles.

3. The development or conversion of software for anthropocentric production is very expensive. We know that in the end, this software will be substantially less capital intensive than its technocentric counterpart, but the changeover from one concept to another is still an expensive venture, for simply providing existing systems with a new user interface won't work in most cases. Many program packages will have to be entirely rewritten to accord with the new role of human beings in the production process. The software is thus an obstacle to change, and the more automated the intellectual work in a factory, and hence the more comprehensive the program systems, the larger the obstacle.

Of course, the fact that even software becomes obsolete tends to counteract this. The development of software technology means that its price/performance ratio is dropping as well, although not so spectacularly as that of hardware, where the earlier escalating maintenance costs are being kept in check too. In view of the grotesque situation existing in most factories whereby first generation programs are run on third generation equipment with second generation operating systems, managers will have to get used to the idea of not only scrapping machines but also consigning programs to the archives in order to clear the way for more efficient systems.

4. The inertia of the social system in the individual company offers more stolid resistance, raising structural obstacles which are difficult to overcome by anthropocentric redesigning (Lutz and Schultz-Wild 1983). With the necessary transfer of essential intellectual work, from production preparation and other centralized technical departments, back into production, it is not only the importance of these departments to company policy that is at issue. The technical employees in these departments are, for the most part, outstanding, skilled workers who have risen there in line with a standard career pattern, and have done it in order to escape the deteriorating working conditions at the production level. The position of these employees is threatened and so they provide a considerable potential for resistance. The position of the foremen is altered too, which could cause them to adopt attitudes ranging from scepticism to downright opposition.

The factory-wide changeover to group technology turns out to be one of great conflict. On the one hand, proceeding in stages enables conflicting principles and forms of organization to exist side by side for a relatively long time, and continually conjures up the danger of reverting to technocentric production. On the other hand, it is the only method which holds out the prospect of success because it brings about change gradually, and allows for experimentation. However, it can only succeed if due regard is given to the long-term, personal plans of employees, offering them clear career prospects and attractive jobs in industry.

5. Finally, the greatest obstacles are to be found in the dogmas of machine-like thought, which have taken hold in the minds of managers, workers and those who represent their interests. Even at the outset, they block out any consideration of alternative forms of manufacturing and work organization, and reject the notion that these could have the necessary rigour. But since new production concepts can only become reality through the minds of people, and can only be applied if they have previously existed as ideas in the brain, human thought has to firstly free itself from its own constraints. It seems that this will best succeed through learning processes initited by being confronted with practical examples of successful conversions.

Beyond this subjective resistance, the relationship with capital places objective barriers in the way of anthropocentric development. It might seem worthwhile from the management's viewpoint to introduce the form of production engineering and labour organization appropriate for the whole workforce, but it can only ever be achieved within certain limits, which take into account managements need to exercise control. In certain cases, the economic advantages of reduced stocks and throughput times and the increased output, quality and availability of machines, may not in fact be swallowed up by the expenditure on additional technical equipment and the higher labour costs.

Moreover, the autonomy of the basic units as opposed to a central planning structure, can only be increased as long as management has no fear of losing control, although of course the actual forms of control will clearly have to change. In technocentric production, control is exercised by the objectivization of knowledge and by the technical apparatus. Here it is exercised by a far more loosely connected computer-based network of central planning specifications and monitoring procedures. Within this network, management "seeks even more strongly than before, to use the judgement, independence and responsibility of the worker as an operational instrument, and because of this, can also count on a higher motivation to work" (Kern and Schumann 1984, p. 175). But motivation, capabilities and the pride of production workers can only develop if suitable conditions prevail in the working environment. The danger is that as a result of the old way of thinking or from lack of courage, these conditions will not be widely appreciated, and so without the scope for action envisaged, potential co-operation will fail to materialize. Companies would then of course, revert to the economic purgatory from which they had hoped in part to escape by using the anthropocentric production concept.

Chapter 5

Horizons New: Farewell to Necessary Work?

Now that the technocentric and anthropocentric production concepts have been counterposed as ideals and as two extremes in the discussion on the shape of future technology, and their respective practical difficulties have been examined, it is time to take stock. How can the two production concepts be economically and politically evaluated, and what are the opportunities for putting them into practice that emerge from this evaluation? Which production concept can be considered the superior one in specific circumstances? What effects will each one have on work, and on prospects for employment? It is the business of this concluding chapter to find provisional answers to these questions.

5.1   Investment Decision Making: Pastures New

It is already clear from the two counterposed courses of development, that under present conditions and those expected in the future, both courses are feasible. Of course, the limits on such developments become quite evident, although in completely differing contexts. Whilst the development of technocentric production is fundamentally limited by essentially incomplete model design, that of anthropocentric production is restricted mainly by the management's need to maintain overall control in a capitalist context. This is quite apart from the fact that the social inertia in the factory of today makes it exceedingly difficult to effect a changeover to this way of working.

In consequence, companies see themselves faced with the choice of taking one path or the other. They cannot compare the technological or economic effects of each one, because information on the two concepts does not exist in such a form. So they are confronted with the difficult task of investigating various possibilities in order to find a production concept which is appropriate for their particular operation. Having to choose however, means they need to have available, suitable criteria for making an evaluation.

The choice proves to be all the more difficult and, at the same time all the more urgent in that unwavering progress along the technocentric route has made it increasingly difficult to turn back, while the anthropocentric route has shown clear economic advantages. Of course, investments made in accordance with one or the other production concept can be judged on economic grounds only by a precise cost–benefit analysis in each individual case. Nevertheless, even when considering the anthropocentric concept in general terms, its enormous potential for efficient restructuring is obvious. Experience of actual cases where production has been restructured in accordance with this concept, shows that in general, apart from the odd exception, throughput times can be shortened extremely effectively, production overheads can clearly be reduced and productivity as well as creativity can be considerably increased. These advantages, which in competition would be decisive, account for the probable economic superiority of this concept, particularly in view of market demands for high flexibility.

As we have seen in Section 2.3.4, the investment calculation procedures available for current use are not really appropriate for assessing the economic merits of the competing production forms. At best, they can be used to evaluate replacement or expansion investments made within the same structural framework, as long as the cash flow situation is fully on record. They are however, bound to fail when attempting to evaluate a qualitative structural change in production, since the accompanying costs and benefits cannot be assigned to particular investment decisions, nor can they be expressed entirely in monetary terms, nor can their life cycle be determined with certainty. It thus proves necessary to find a way of making reliable estimates of these aspects.

Now, the transition from technocentric to anthropocentric production, represents exactly that kind of qualitative leap which throws up these evaluation difficulties. Consequently, new tools for assessment are urgently required in order to evaluate both those investments which bring about structural changes made in the transition from one concept of production to another, and those which finance production systems planned according to concepts other than the present ones, together with their compatibility with current or future production structures. System and structure comparisons of this type are hardly carried out at all, even though they are urgently necessary in view of permanent changes in the world market, and the accompanying demand for new production structures. Instead of this, investments are quite overwhelmingly evaluated according to conventional procedures based on existing structures, in the same way as before, with no consideration given to alternatives which might be more productive and profitable. In addition, the productive potential of human labour on which such things as rapid reorganization, high systems utilization and swift innovation decisively depend, remains systematically excluded from the assessments which only refer to human beings in the ugly old guise of labour costs.

As we shall now show, this one-sided retention of existing production structures and the low receptivity for fundamentally different concepts can lead to surprising economic effects. Technocentric investments which might be highly profitable in the short term can prove to be the less profitable alternative in the more long term. As the neglected capabilities of human labour gradually waste away and are lost as a productive asset, so in work which has been objectivized, every specialization restricts the potential for a qualitative change in the procedures. A production system dedicated to a definite, very narrow range of parts can consequently forfeit its profitability in comparison with an alternative system which may be less productive at first, but has a more universal design, and can be harnessed by human beings as soon as significant changes are required in the range of parts. In this alternative system, flexibility, high utilization and creativity are preserved. Many companies have gone bankrupt by making investments that are profitable only in the short term, because they deployed a system which, in the final analysis, was unsuitable for production and which could no longer satisfy the market.

The following sample calculation on investments in manufacturing based on different production concepts may clarify this state of affairs in more detail (CSS 1981, pp. 78f). A dynamic treatment of the production process over a long period of time, enables a series of points in time to be established at each of which, a decision is taken on investments which alter the technology and labour organization of the process. For the sake of simplicity, it may be assumed that at each decision point, a choice is made between alternative investments which require the same capital expenditure but lead to highly different production structures. In complete accordance with the counterposed production concepts we have been discussing, we will denote with T the one structured on a Taylorist basis, and the one structured on a non-Taylorist basis we will denote with N. These structures are retained until a choice is once again made between these two alternatives at the instant of the next investment. After several decision points where the choices are made, a sequence of decision results is obtained, for example, the sequence $TTNN$ is a possible outcome.

The alternative decision results bring about different economic effects. It may be assumed that the Taylorist structure represents an increase in productivity which is twice as big as that of the non-Taylorist structure. For example, an increase of 100% in the case of a T decision [an increase by the factor $(1 + 1)$] instead of only 50% in the case of an N decision [an increase by the factor $(1 + 0.5)$]. In a Taylorist production structure of course, the skills and abilities of human beings wither away and at later stages are no longer available. In the model, this fact is denoted by a factor b, which reduces the increase in productivity in the period following a T decision, for example to one fifth if only 20% of the labour force were still able to retain their productive abilities. At the end of the sequence TT therefore, there would have been a productivity increase by the factor 2×1.2, in which the factor 2 is due to the first T decision and the factor of only 1.2 is due to the second T decision. In accordance with these assumptions, the sequence TN would increase productivity by the factor 2×1.1 the sequence NT by the factor 1.5×2 and the sequence NN by the factor 1.5×1.5. In formal notation, the productivity P after n steps is as follows:

$$P_n = P_0 \prod_{i=1}^{n} (1 + a_i b^k)$$

where P_0 is the initial productivity

P_n is the productivity after n decisions

a and b are as described above

k is the number of T decisions already made

In the examples given:

$a = 1.0$ for a T decision

$a = 0.5$ for an N decision

$b = 0.2$

By way of example, Fig. 21 shows the productivity increase that would result for all possible sequences of four decision steps using the above numerical values.

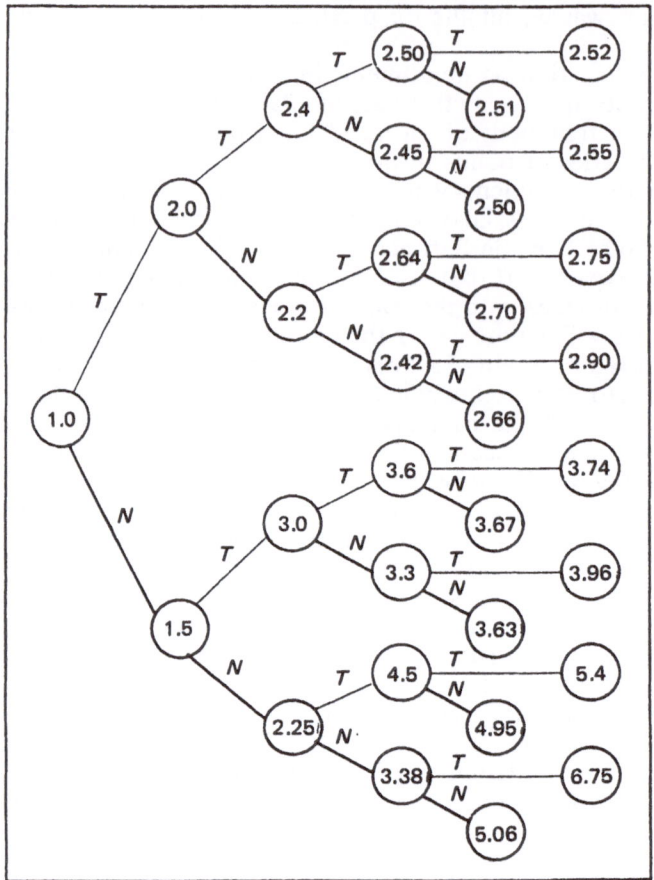

Fig. 21.   Productivity increases resulting from different investment strategies.

Comparing the sequence $TTTT$ with the sequence $TNNN$ shows that in the second sequence the productivity is increased by a factor of 2.66 overall, compared with only 2.52 in the first sequence, even though the individual rates of increase from N decisions are only half as bigh as those from T decisions. If N had been chosen in the first decision as well, productivity would have risen by a factor of 5.06 overall. As the structure of the model shows, this effect occurs in every case, even if different numerical values are used so long as a sufficiently long sequence of N decisions can be kept up under conditions of competition. The effect is essentially based on the fact that, in accordance with findings in industrial psychology, the productivity of human labour gradually disappears in Taylorist production structures, while in non-Taylorist structures it can at least be maintained.

   This simple model calculation serves to demonstrate the great influence of production structures on the economics of an investment, and how important it is for investment decisions to be linked up with a long term investment policy.

This does not of course mean that monetary evaluation of investments is in any way obsolete. The most productive solution with the lowest unit costs is without question the Taylorist one if the firm specializes in only a few parts which do not change during the lifetime of the investment. As soon as this becomes uncertain, a production system capable of more universal application seems to be the more profitable investment overall, even though the initial result is lower output and higher unit costs. For it is equally true, that flexibility is nothing other than surplus capacity, qualitative as well as quantitative, which can be looked upon as an insurance against uncertain production demands at a future stage.

Working out a production structure to suit market demands and then investing in its implementation, necessitates far-reaching economic assessments and policy decisions on the part of the company. It will not be enough to just rationalize parts of the production process according to traditional recipes so long as the production process as a whole is insufficiently suited to the new world market situation. On the contrary, clinging on to conventional methods of assessing investment, means that the accountant becomes the real boss in the factory, or, to be more precise, the investment calculation program. Neglecting to make the necessary economic assessments and crucial company policy decisions, can of course mean that before long, there will be no more business left.

5.2   Work Planning Is Available in Every Case

Work scheduling is an area of industrial staff work which is very expensive but which can however, proceed according to precise rules, even though they have to be specified for each company. Lowering the costs of this work, systematizing and objectivizing it arouses a twofold interest on the part of management in automating it through the use of computers, for apart from these advantages, it can be carried out in a manner which is entirely independent from the knowledge of an individual specialist. There are of course, widely differing technological and organizational options for getting this done.

The use of computers along technocentric lines ends up with complete automation of the work scheduling tasks which are not associated with specific customer orders, at least for limited classes of parts. The inputs required for this are restricted to the formal description of the workpiece. Where inadequate modelling does not permit this, the work schedules and NC programs are drawn up interactively, using the computer. For example, the extent of the parts range, and the variety of production methods and machine tools often call for decisions which are dependent on the situation. In this event, sequences of work operations, machines, tools and fixtures used have to be input along with the workpiece description.

In this type of computer facility, the operation sequence, which previously would have been determined by the work scheduler, is specified by the scheduling system incorporated in the computer. So now the human scheduler has become subordinate to the system. The new working practices are firmly formalized and routine, being linked to the sequences determined by the computer system. It is no longer the work scheduler but the computer program which fixes the "optimum" production operation; for the planning algorithms which have been implemented, merely require situation-specific inputs in order to determine these operations. The work scheduler in the technical office, like the skilled worker in the workshop before him, now sees himself subjugated to the compulsory, largely automatic machine sequence. He now finds himself in the exact situation he had hoped to escape from, when he followed a well established career pattern and moved up from the workshop.

The working conditions in the technical office are now deteriorating in the same way as they did in the workshop, being a consequence of the division of labour, formal working and the use of machines operated on the same rationalization principles. They make the career rise into the office appear progressively less attractive to the bright skilled worker who is conscious of his knowledge and capabilities in the area of production, just as they sour the supposed promotion of those who have already moved there. As a result, work scheduling is no longer an attractive option within the company career structure, and as a career aim, it has largely had its day.

This is just in work scheduling, but as further progress is made along the technocentric route, the goal of computer integrated manufacture is just around the corner. This would link the CAP system in work preparation with the CAD system in design, and thus ensure the internal transfer of geometrical data structures. This having been done, formal workpiece descriptions, would no longer be necessary. The extent to which planning knowledge can be formalized and planning operations can be controlled, makes the use of expert systems to further automate work scheduling seem highly promising. This would not however, in any way, enable the work situation to take a turn for the better.

In comparison, anthropocentric production in the production island clearly exhibits more advantageous working conditions. Planning and implementation are largely reunited in the complete production of parts families. For the skilled worker, this reintroduces a comprehensive work procedure, presents him with a challenge and does not fuel him with a desire for promotion to the supposedly superior technical office. For the specialist worker in work scheduling, the production island likewise offers a meaningful perspective, for in this case, production procedures are planned less according to specified rules and far more according to the situation at hand, being directly connected to production itself. So it is that the work scheduler can again be of assistance, and can renew his own store of personal experiences through exchanges with the skilled worker.

In view of the capabilities required, and the fields of activity covered in anthropocentric production islands, the teams working there can be comprised of both former workshop personnel and former work schedulers, just as in flexible manufacturing systems, and in production centres with integrated organization and planning tasks. Both the fading attractiveness of work scheduling and this positive prospect of work in production islands, will decisively weaken the opposition of work schedulers to the transfer of certain areas of authority back into production. Consequently, the principles of group technology present a strong technological and organizational basis for a compromise solution to the conflict of interests that is to be expected between skilled workers and work schedulers. They could also remove a substantial obstacle to anthropocentric development by relieving the works councils of their dilemma on the issue of representation.

Bearing all this in mind, traditional work scheduling is bound to disappear as an essential part of a separate production preparation function. This will happen either through almost total automation, or by work scheduling being re-located again and combined with implementation. There are of course further suggestions along the lines of maintaining the division of planning and implementation, by forging a strong link between computer-based work scheduling (including NC programming) and design, and possibly incorporating work scheduling into the design function in an organizational sense (RKW 1984). In view of the many insuperable planning deficiencies, due mainly to the failure to take account of current production conditions and the work situation, this organizational link with design must appear just as dubious as the internal transfer of geometrical data structures within an integrated computer system. Since these planning deficiencies can only be reduced effectively if planning and implementation are reintegrated, the transfer of work scheduling back into production holds the promise of much greater efficiency.

The automation of intellectual work in production preparation entails new tasks and lends greater importance to those which were previously peripheral. New tasks arise from the development, introduction and maintenance of the program systems used for work scheduling and NC programming. As far as manufacturing is concerned, the main jobs are to work out the functional demands on the CIM modules, to introduce the computer systems, to maintain their programs and to adapt them to new conditions. The nature and cost of these activities differ considerably of course, depending on which development route is followed.

Technocentric production requires enormous expenditure on the introduction and maintenance of the CIM modules, because the production process, as we have seen, must be completely modelled in the CAP system, and each change in method, machines, and means of production means altering the data structures and programs. With anthropocentric production on the other hand, costs are limited to whatever is required for strategic planning. Only general work schedules, independent of customer orders are programmed; the programming routines for human interaction with the system have to be worked out and maintained. The operational planning of the production process within these general conditions, is then a matter for the teams in the production islands. This means though, that general, strategic planning becomes a task of the utmost importance from management's viewpoint because the specifications laid down by it, together with the continuous monitoring of the state of production, are the only things that management can use to maintain control in the factory and to regulate output, quality and costs.

Technology investment planning now takes on a role of far greater importance. Its role is to plan production methods and procedures from a medium-term perspective, to work out from this the functional requirements of the means of production to be acquired (machines and program systems) and to co-ordinate their capabilities. In view of the conclusion drawn in the previous section, it is obviously essential to link technology investment planning with company strategy and the production concept on which it is based. In all this, the primary concern is to establish the principles of work organization, technological development and personnel development which have to serve as strict specifications for planning technological investment. In the absence of this exercise, investment policy would be governed by technocratic criteria alone, and the primacy of technology would remain unchallenged.

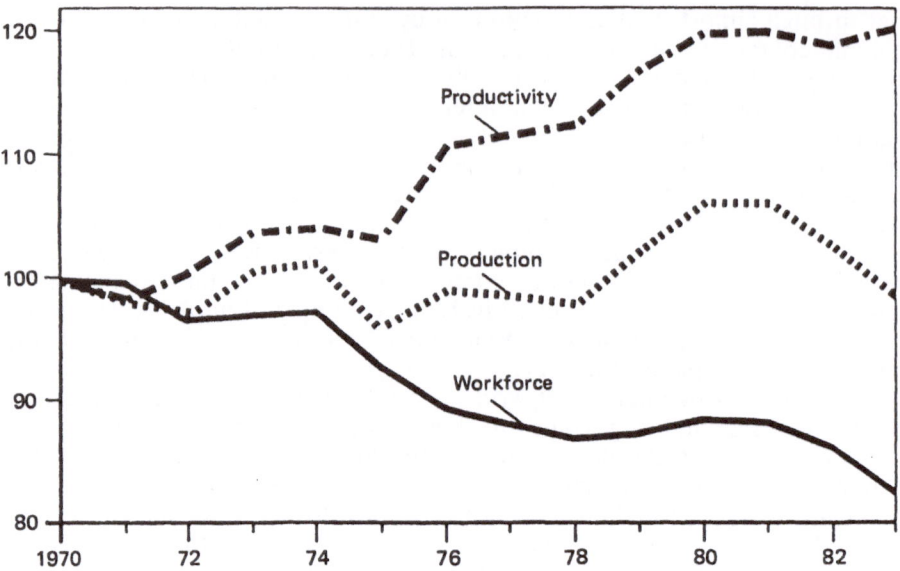

Fig. 22. Workforce, production and productivity in the mechanical engineering industry (not including office machinery and general data-processing), 1970 = 100. Source: Statistisches Bundesamt.

5.3   The Dangers of Segmentation

Technological and organizational change will bring about redundancies on a large scale, in both courses of development, though in different ways and to differing extents. Since the beginning of the 1970s, there have been clear signs of stagnation in the capital-goods industry. Rises in productivity result in a drop in employment when there is no corresponding rise in production. Bearing in mind that mechanical engineering (without office machines or data-processing equipment) is the most important sector of this industry, a worrying picture emerges. Despite export surpluses, which were at a level of about DM 50 000 million in 1982, as can be seen from Fig. 22, production in 1983 was at the same level as in 1970 whilst productivity during this period increased by 20% and employment fell by 17%.

For the reasons discussed in Section 2.3.5, there is no prospect of growth in production for a long time to come. It will be much more a case of "fighting your own corner" in a situation of intensified competition, where the volume of production remains essentially the same and where the object is to knock out your opponents. If the technocentric course is followed, the rate of redundancies will increase, and this is quite separate from cyclical unemployment. This assessment is based on three considerations in particular.

Firstly, we are in the diffusion phase of CNC machines, where their rate of spread is at its highest, so their use will increase still further throughout the whole range of customer-led (order-bound) small scale production. On average, a CNC machine can offer a level of output three to four times higher than that of a conventional machine tool. Multiplied by the figure of about 5000, which is the number of units added each year, this innovation leads in the course of time, to a noticeable reduction in the number of production units, which require correspondingly fewer operating personnel.

Secondly, owing to automatic monitoring devices, the periods during which these capital-intensive devices can be used without operating personnel will increase. Even if complete ghost shifts are not yet likely, the fully automatic operation of these devices, occurring from time to time during breaks and during parts of the second or third shift, will nevertheless contribute noticeably to an increased level of redundancy.

Thirdly, it can be expected that staff work in the technical offices will become extensively automated in the near future, beyond areas of experimentation to company information processing in the form of individual restructured areas.

Although its overall share of the work will continue to increase overall, office redundancies will have to be reckoned with. Areas of work particularly hard hit will be drawing, work scheduling and parts lists organization, as well as materials procurement and production control. A complete and comprehensive integration of all the automated individual processes in the computer network (CIM) on the other hand, will still have to wait, owing to the tremendous number and complexity of the interfacing problems that still have to be solved. A move in this direction, would of course, bring about even higher savings on overall labour costs.

If the anthropocentric route is followed, substantially greater redundancies can be expected as a result, for if it is pursued logically, the rate of increase in productivity will be higher, for reasons already discussed. Firstly, we must assume that CNC machines and monitoring devices will be used for roughly comparable amounts of work. That being the case, the noticeably higher increase in productivity can then be attributed to two factors which are connected with the comprehensive reintegration of planning and implementation. The expenditure on work scheduling, NC programming and organizational tasks can be reduced and at the same time their quality can be increased. Time lost in correcting and optimizing imperfect schedules and programs can largely be avoided. If the experiential knowledge of the workers is used directly, the work schedules and organizational decisions correspond much more precisely with the current state of production, whilst the NC programs have shorter running times. Even though a comparatively smaller number of functions of human labour are automated in anthropocentric production, given the number of functions that are appropriate for human abilities and those requiring human/machine interaction, the possible gain in productivity is still substantially higher. The potential existing here for long term productivity has already been researched, and the possible reduction in the number of indirect producers for every 100 direct producers from 144 to 90 seems feasible.

Technical employees are the ones primarily affected by these additional redundancies. This explains why the redundancies in the capital goods industry, which have come about as a result of technological and organizational change, will carry on increasing, with no prospect of compensating effects through higher production for example. Not only that, if anthropocentric development is carried through, redundancies will be higher still. Ironically therefore, the technocentric concept seems to offer the greater "protection against rationalization" at the moment. But appearances can be deceptive. The smugness of managements

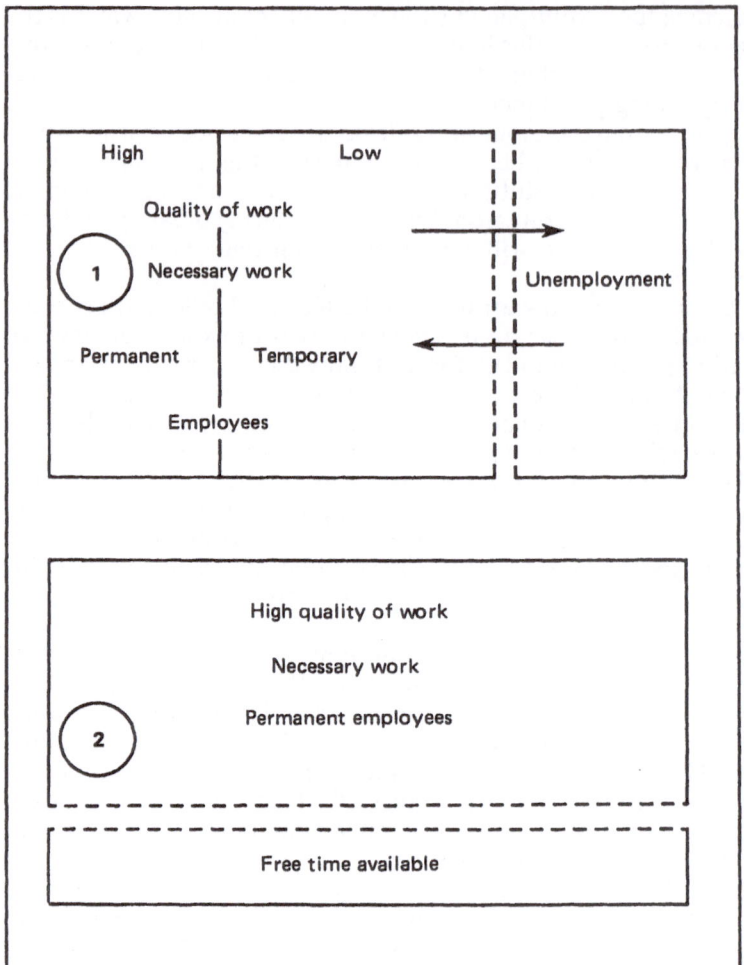

Fig. 23.   Different allocations of the volume of work.

misled by conventional, efficiency comparisons, and the immobility of works councils led astray by their fierce defence of traditional social structures in the company, can at best, guarantee only temporary protection. The gains in rationalization that can be obtained by creeping along the technocentric route like this can well turn out to be too small to maintain competitiveness indefinitely, and redundancies avoided in the short term can very quickly turn into no jobs at all in the long term.

As we have seen earlier, the anthropocentric production concept makes considerable demands on skilled workers. They need broad specialist knowledge and abilities not related to a particular function. It is because of this and the higher potential for productivity that it clearly runs the risk of causing a split in the workforce, both inside and outside the factory, into highly qualified permanent employees with favourable working conditions, and less qualified workers who are constantly being threatened with unemployment (as shown in the upper part of Fig. 23). This risk can be attributed to the following set of circumstances. Given the present situation in the labour market, and without counter-measures being taken, unemployment initially rises. For obvious reasons, companies then tend to retain the better qualified and capable workers at every stage, gaining support from the anthropocentric concept itself, which requires a high degree of employee capability. These workers can thus be made highly productive. For less well qualified workers however, these outstanding requirements could prove to be an insurmountable obstacle to finding work again or to changing jobs within anthropocentric production areas (Kern and Schumann 1984).

A labour market segmented in this way, would provide management with the opportunity to impose sanctions in order to ensure control over the workforce. The principle of "divide and rule" employed so far, could now be replaced with the "carrot and stick" principle for controlling the workers. Referring to the top of Fig. 23, the "carrot" is the move from an insecure area into an area of highly skilled permanent employment, whilst the "stick" is the slide into insecure employment or even unemployment. In order that the sanctions can be effective, these two aspects must not be entirely separated, but should have boundaries that are weak enough to be crossed now and again.

The tendency to segment the workforce and the risks involved, are not of course an inevitable consequence of the spread of anthropocentric production. It is quite possible for the transition from the Taylorist factory to the factory with a skilled workforce, to be carried out in such a way that this split is avoided and not promoted. From the viewpoint of quantity, the inevitable reduction in the volume of necessary work, must be offset by a general reduction in working hours. It is quite possible to finance this move, because of the extraordinary increases in productivity. From the viewpoint of quality, anthropocentric production requires broad qualifications and capabilities which enable as many workers as possible to show that they are up to the comparatively high standard necessary. Part of this demand includes new skills, such as thinking and acting in a systems context or in co-operation with others. In Germany, these skills can be developed to a level which is very high by international standards, but the procedures have to allow for different learning methods according to initial skills. For example, semi-skilled workers would learn in a different way to skilled workers.

IG Metall's new campaign programme "Work and Technology", is quite right to place particular emphasis on skill acquisition measures of this type (IG Metall 1984b). In a recent agreement which has revised the job structure in the metal-working industry, and which sets out to impact a variety of skills and knowledge and avoid one-sided specializations, a training basis has been set up which will serve to disseminate the anthropocentric concept. With a consistent policy of negotiating for a shorter working week and broader qualifications (with controls on performance), a distribution of essential work such as that shown at the bottom of Fig. 23 can definitely be arranged, providing the anthropocentric concept is pursued. Given the shorter working week, this distribution provides skilled work for as many workers as require it, with permanent employment in conditions suitable for human beings.

Achieving this distribution and organization of essential work in such a way, presupposes adequate social conditions, which of course can only come about if production workers articulate an interest in getting work designed so that it is suitable for human beings. Workers have to use their powers of representation in the union and in their company to influence technological and organizational development themselves. If they do not, the new kind of sanctions and burdens which have been discussed earlier, would make headway, for the anthropocentric concept can't just of itself create the working conditions suitable for human beings in general.

5.4   Conclusion: For and Against the Anthropocentric Production Concept

The progress of rationalization in the context of capitalism has taken a remarkably spiral-shaped course. This progress began with the political problem of keeping control, and the solution for this was the technological division of labour and the use of machines. At the same time, this solution proved to be highly efficient economically. In production based on the division of labour, the demands for objectivizing knowledge, the control of human labour and the production of relative surplus value are initially in harmony. In the course of events, the increasingly automated production process tends to remain stuck with its programmed sequences, whilst the procedure upon which this programming relies, that of objectivizing human labour, comes up against barriers as well. The production economy and the market, come into conflict with its concern for exploitation and control. Overcoming this conflict requires new production concepts which give back to human labour the main role in production, no longer in terms of quantity admittedly, but hopefully in terms of quality. Technology and economics have become political again.

Despite all the problems which arise if the anthropocentric production concept is pursued to its logical conclusion, it can as we have seen, lay claim to economic superiority for several reasons. In competition in the world capital goods market, there are weighty arguments in its favour regarding flexibility, quality and prompt delivery. Even when the reorganization of production to fit this concept results in major problems of a social nature in addition to those of a technological and organizational nature, they can all be overcome. With technocentric production, competitiveness could at best be maintained, but improving it probably wouldn't be possible. The company could not indefinitely avoid the fact that it would cause crucial productive and creative abilities to wither away and with that, a loss in the power to innovate. In contrast, anthropocentric production holds out the promise of further progress in using as resources, those qualifications and capabilities that are available in comparative abundance in the Federal Republic of Germany.

Apart from economic considerations, the choice between different courses of development is of vital significance in the way that human beings live. With technocentric production, characterized by automation, the production and accumulation of surplus value may indeed justify the technology, but at the same time, the human being's role as the subject of history would be reduced. Insofar as he is occupied at all in production, he is subjected to mechanistic working conditions that force him to act and think like a machine himself.

The most devastating outcome of automation is therefore the fact that its end-product is the auto-mated organizational human being, the person who receives his instructions from the system and, as a scientist, engineer, specialist, administrator or ultimately as consumer and citizen, cannot imagine himself deviating from the system, not even in the interests of efficiency, and even less so in order to find a more reasonable, more lively, more useful and more decent life-style. (Mumford 1977, p. 555)

The outcome of this, is that spurred on by necessities, the human being is compelled to behave in a way that robs him of his ability to develop in the realm of freedom.

In contrast, anthropocentric production would reduce the number of compulsory activities sharply, much more so than would otherwise be possible, even under the restrictions of competition. At the same time, it would create conditions whereby the human being remaining could make use of the skills already acquired and could develop new abilities. Assuming the existence of adequate related conditions in accordance with negotiating policy, this concept would ensure that compulsory work procedures did not increase any more than necessary, and that in all areas, both manual and intellectual, the work itself would become a challenge for those with production knowledge and capabilities. Surely this is sufficient reason to remove the impediment to anthropocentric production.

Bibliography

Agurèn S. and Edgren, J. 1983 "Neue Wege der Produktions- und Fabrikplanung", Rationalisierungskuratorium der deutschen Wirtschaft e.V., Eschborn

Ahlmann, H.-J. 1980 "Fertigungsinseln – eine alternative Produktionsstruktur", *Werkstatt und Betrieb* **113** pp. 641–648

Altmann, N. and Bechtle, G. 1971 *Betriebliche Herrschaftsstruktur und industrielle Gesellschaft. Ein Ansatz zur Analyse*, Munich

Altmann, N., Bechtle, G. and Lutz, B. 1978 *Betrieb – Technik – Arbeit. Elemente einer soziologischen Analytik technisch-organisatorischer Veränderungen*, Frankfurt/New York

Anders, G. 1973 *Die Antiquiertheit des Menschen*, vol. 1, Munich

Autorengemeinschaft 1983 "CAD im Maschinenbau. Diskussionsergebnisse aus dem VDMA, insbesondere aus den Fachkreisen 'TB-Organisation'", *Technische Zeitschrift für praktische Metallbearbeitung* **77** pp. 33–38

Autorengemeinschaft (ISI, IAB, IWF) 1982 "Der Einsatz flexibler Fertigungssysteme. Technische, einführungsorganisatorische, wirtschaftliche und arbeitsplatzbezogene Aspekte", Kernforschungszentrum Karlsruhe, KfK-PFT 41

Autorengemeinschaft 1984 "Autonome Fertigungsinsel", Kernforschungszentrum Karlsruhe, KfK-PFT 79

AWF (ed.) 1984 *Flexible Fertigungsorganisation am Bespiel von Fertigungsinseln*, Ausschuss für wissenschaftliche Fertigung e.V., Eschborn

AWK 1984a *Autorenkollektiv* "Einsatz von CAD-Systemen", *Vortrag zum Aachener Werkzeugmaschinen-Kolloquium*

AWK 1984b *Autorenkollektiv* "Wandel der Arbeitsplanung bei EDV-Einsatz", *Vortrag zum Aachener Werkzeugmaschinen-Kolloquium*

AWK 1984c *Autorenkollektiv* "Marktorientierte Produktionsplanung und -steuerung", *Vortrag zum Aachener Werkzeugmaschinen-Kolloquium*

AWK 1984d *Autorenkollektiv* " Flexible Fertigungsanlagen", *Vortrag zum Aachener Werkzeugmaschinen-Kolloquium*

AWK 1984e *Autorenkollektiv* "Automatische Fertigungsüberwachung", *Vortrag zum Aachener Werkzeugmaschinen-Kolloquium*

Babbage, C. 1835 *On the Economy of Machinery and Manufactures*, London

Bahr, H.-D. 1983 *Über den Umgang mit Maschinen*, Tübingen

Bailey, H. 1983 *Job Design and Work Organisation*, Englewood Cliffs/London/New Delhi/Rio de Janeiro/Singapore/Sydney/Tokyo/Toronto/Wellington

Bainbridge, L. 1982 "Ironies of automation", *Proceedings of the IFAC/IFIP/IFORS/IEA Conference on Analysis, Design, and Evaluation of Man–Machine Systems*, Oxford/New York/Toronto/Sydney/Paris/Frankfurt

Bammé, A., Feuerstein, G., Genth, R., Holling, E., Kahle, R. and Kempin, P. 1983 *Maschinen-Menschen, Mensch-Maschinen. Grundisse einer sozialen Beziehung*, Reinbek

Baumgartner, P. and Stöckert, W. 1984 "CAD-Einsatz in Europa, Japan und den USA", *Werkstatt und Betrieb* **117** pp. 197–203

Bechmann, G., Huxdorff, K., Vahrenkamp, R., Werle, R. and Wingert, B. 1978 "Auswirkungen des Einsatzes informationsverarbeitender Technologien untersucht am Bespiel von Verfahren des rechnergestützten Kontruierens und Fertigens (CAD/CAM)", Kernforschungszentrum Karlsruhe, KfK-CAD 114

Bednarz, K., Heitmann, G. and Kempin, P. 1984 *CAD/CAM und Qualifikation*, Frankfurt/New York

Beitz, W. 1983 "Entwicklungszwänge für den Konstruktionsprozess", *Produktionstechnisches Kolloquium Berlin, PTK '83*, Munich

Benz-Overhage, K., Brumlop, E., von Freyberg, T. and Papadimitriou, Z. 1982 *Neue Technologien und alternative Arbeitsgestaltung*, Frankfurt/New York

Bittcher, A. 1981 "Die Fertigunsinsel als Erscheinungsform einer dezentralen Betriebsorganisation", *Werkstatt und Betrieb* **114** pp. 147–151

Bjørke, O. 1983 "Geometrisches Modellieren als integraler Bestandteil rechnergestützter Konstruktions- und Fertigungssysteme", *Produktionstechnisches Kolloquium Berlin, PTK '83*, Munich

Blauner, R. 1964 *Alienation and Freedom*, Chicago/London

Boberg, J., Fichter, T. and Gillen, E. (eds.) 1984 *Exerzierfeld der Moderne. Industrie-Kultur in Berlin im 19. Jahrhundert*, Munich

Brankamp, K. 1981 "Gesamtauftragssteuerung in Unternehmen der Einzel- und Kleinserienfertigung", *VDI-Zeitschrift* **123** pp. 35–40

Braverman, H. 1977 *Die Arbeit im modernen Produktionsprozess*, Frankfurt/New York

Bright, I.R. 1959 *Automation and Management*, Boston

Brödner, P. 1984 "Group technology – a strategy towards higher quality of working life" in T. Martin (ed.) *Design of Work in Automated Manufacturing Systems*, Oxford/New York/Toronto/Sydney/Paris/Frankfurt

Brödner, P. and Hamke, F. 1969, 1970 "Automatisierung und Arbeitsplatzstrukturen", *Mitteilungen des IAB* pp. 604–618 (1969), pp. 137–172 (1970)

Brödner, P., Krüger, D. and Senf, B. 1981 *Der programmierte Kopf*, Berlin

CEC 1984 "ESPRIT 1984 Workplan", *Official Journal of the European Communities*, Commission of the European Communities, **27** C47/1–67

Cooley, M. 1978 *Computer Aided Design – sein Wesen und seine Zusammenhänge*, Stuttgart

CSS 1981 "New technology: society, employment and skill" *CSS Report*, Council for Science and Society, London

Daly, A., Hitchens, D. and Wagner, K. 1985 "Productivity, machinery and skills in a sample of British and German manufacturing plants", *National Institue Economic Review* pp. 48–61

d'Iribarne, A. 1982 "Flexible Fertigungssysteme in Japan: technische, wirtschaftliche und soziale Aspekte" in B. Lutz & R. Schultz-Wild (eds.) *Flexible Fertigungssysteme und Personalwirtschaft*, Frankfurt/New York

Dietz, P. 1983 "Baukastensystematik und methodisches Konstruiren im Werkzeugmaschinenbau", *Werkstatt und Betrieb* **116** pp. 185–240

Dietzgen, J. 1930 "Das Wesen der menschlichen Kopfarbeit" in E. Dietzgen (ed.) *Sämtliche Schriften*, Berlin

Dörr, G., Hildebrandt, E. and Seltz, R. 1984 "Kontrolle durch Informationstechnologien in Gesellschaft und Betrieb" in U. Jürgens & F. Naschold (eds.) *Arbeitspolitik*, Opladen

Dolezalek, C.M. and Rophohl, G. 1970 "Flexible Fertigungssysteme – die Zukunft der Fertigungstechnik", *Werkstattechnik-Zeitschrift für industrielle Fertigung* **60** pp. 446–451

Dreyfus, H.L. 1979 *What Computers Can't Do. The Limits of Artificial Intelligence*, New York

Dreyfus, H.L. and Dreyfus, S.E. 1986 *Mind over Machine*, New York

Edwards, R. 1981 *Herrschaft im modernen Produktionsprozess*, Frankfurt/New York

Eigner, M. 1983 "Anforderungen und Voraussetzungen für die betriebliche Integration von CAD-Systemen", *VDI-Zeitschrift* **125** pp. 187–194

Eigner, M. and Maier, H. 1983 "CAD-Marktübersicht", *Fortschrittliche Betriebsführung/Industrial Engineering* **32** pp. 234–249

Eversheim, W., Dahl, B. and Schulze, P. 1984 "Integrierte Erstellung von Fertigungsunterlagen", *Computer Magazin* **9** pp. 32–36

Feigenbaum, E.A. and McCorduck, P. 1983 *The Fifth Generation*, Reading

Friedrichs, G. and Schaff, A. (eds.) 1984 *Auf Gedeih und Verderb*, Reinbek

Fröhner, K.-D. and Duda, S. 1979 "Steuerungssysteme für Werkstattfertigung und Handlungsmöglichkeiten betrieblicher Mitarbeiter", *Fortschrittliche Betriebsführung/Industrial Engineering* **28** pp. 39–45

Gauderon, E. 1983 "Fertigen in einer autonomen Fertigungsinsel", *Werkstattstechnik-Zeitschrift für industrielle Fertigung* **73** pp. 739–741

Genschow, H. 1983 "Bearbeitungszentren als Bausteine flexibler Fertigungszellen", *VDI-Zeitschrift* **125** pp. M3–M7

Gerlach, H.-H. and Vortherms, B. 1977 "Probleme beim Einsatz von EDV-Systemen in der Produktionssteuerung", *Werkstattstechnik-Zeitschrift für industrielle Fertigung* **67** pp. 629–634

Gerwin, D. 1982 "Arbeitnehmerreaktionen auf flexible Fertigungssysteme und Folgerungen für die Arbeitsorganisation" in B. Lutz & R. Schultz-Wild (eds.) *Flexible Fertigungssysteme und Personalwirtschaft*, Frankfurt/New York

Giedion, S. 1982 *Die Herrschaft der Mechanisierung*, Frankfurt

Gorz, A. 1980 *Abschied vom Proletariat*, Frankfurt

Grabowski, H. 1982 "Schlüsselfertige CAD-Systeme – technischer Stand", *VDI-Zeitschrift* **124** pp. M57–M76

Grabowski, H. and Eigner, M. 1980 "Auswahl und Einführung von schlüsselfertigen CAD-Systemen", *Fortschrittliche Betriebsführung/Industrial Engineering* **29** pp. 149–162

Hacker, W. 1978 *Allgemeine Arbeits- und Ingenieurpsychologie*, Bern/Stuttgart/Vienna

Ham, I. 1977 "Group technology applications for computer aided manufacturing", *NC/CAM Journal* pp. 17–22

Hammer, H. 1983a "Verbesserung der Wirtschaftlichkeit durch flexible Automatisierung beim Bohren und Fräsen", *Zeitschrift für Wirtschaftliche Fertigung* **78** pp. 77–86

Hammer, H. 1983b "Neue Lösungen zur flexiblen Automatisierung der spanenden Bearbeitung", *Produktionstechnisches Kolloquium Berlin, PTK '83*, Munich

Hatvany, J. 1983 "CAD – gegenwärtiger Stand und Ausblick", *Produktionstechnisches Kolloquium Berlin, PTK '83*, Munich

Hatvany, J. (ed.) 1983 *World Survey of CAM*, Sevenoaks

Hellwig, U., Hellwig, H.-E. and Paulus, M. 1983 "Die Kopplung von CAD und CAM", *VDI-Zeitschrift* **125** pp. 355–360 (Teil 1), pp. 455–460 (Teil 2)

Hesser, W. and Rybak, H. 1979 "Rechnergestütztes Konstruieren in Grenzen", *VDI-Nachrichten* no. 37

Hirsch-Kreinsen, H. 1982 "Folgen der EDV-Fertingungssteuerung in Maschinenbaubetrieben", *REFA-Nachrichten* pp. 197–201

Hirsch-Kreinsen, H. 1984 *Organisation mit EDV*, Frankfurt

Hörl, A. 1982 "Flexibles Fertigungssystem fur scheibenförmige Rotationsteile", *Werkstattstechnik-Zeitschrift für industrielle Fertigung* **72** pp. 9–13

Holzkamp, K. 1978 "Sinnliche Erkenntnis – historischer Ursprung und gesellschaftliche Funktion der Wahrnehmung, Königstein/Ts.

IFO 1982 *ifo-schnelldienst* nos. 17, 18, 24, 25, ifo Institut für Wirtschaftsforschung, Munich

IG Metall (eds.) 1983 *Bildungsbaustein CAD/CAM*, Frankfurt

IG Metall (eds.) 1984a "Einsatz und Auswirkungen neuer Technologien in Planung und Konstruktion CAD/CAM", *Arbeitsheft 5 der IG Metall zur Humanisierung des Arbeitslebens*, Frankfurt

IG Metall (eds.) 1984b *Aktionsprogramm: Arbeit und Technik – der Mensch muss bleiben!*, Frankfurt

Jakob, W. 1981 "Fertigung kundenspezifischen Zubehörs in einer Fertigungsinsel", *Werkstatt und Betrieb* **114** pp. 9–12

Janzen, K.-H. 1980 "Alte Aufgaben mit neuen Problemen", *Der Gewerkschafter* **5** p. 2

Janzen, K.-H. 1983 "Menschenleere Fabrik – Fortschritt für wen?", *Produktionstechnisches Kolloquium Berlin, PTK '83*, Munich

Jones, B. 1984 "Division of labour and distribution of tacit knowledge in the automation of metal machinery" in T. Martin (ed.) *Design of Work in Automated Manufacturing Systems*, Oxford/New York/Toronto/Sydney/Paris/Frankfurt

Jürgens, U. 1984 "Die Entwicklung von Macht, Herrschaft und Kontrolle im Betrieb als politischer Prozess – eine Problemskizze zur Arbietspolitik" in U. Jürgens & F. Naschold (eds.) *Arbeitspolitik*, Opladen

Keil-Slawik, R. 1985 "KOSMOS. Ein Konstruktionsschema zur Modellierung offener Systeme als Hilfsmittel für eine ökologisch orientierte Softwaretechnik", dissertation, TU Berlin

Kern, H. and Schumann, M. 1977 *Industriearbeit und Arbeiterbewusstsein*, Frankfurt

Kern, H. and Schumann, M. 1984 *Das Ende der Arbeitsteilung? Rationalisierung in der industriellen Produktion*, Munich

Kettner, H. and Bechte, W. 1981 "Neue Wege der Fertigungssteuerung durch belastungsorientierte Auftragsfreigabe", *VDI-Zeitschrift* **123** pp. 459–465

Kocka, J. 1969 *Unternehmensverwaltung und Angestellenschaft*, Stuttgart

Köhler, C. and Schultz-Wild, R. 1983 "Flexible manufacturing systems – manpower problems and policies", *Proceedings of World Congress on the Human Aspects of Automation*, Ann Arbor

Kölle, J., Döbele-Berger, C., Martin, H. and Martin, P. 1984 "Entwicklung des Informationssystems PSK 2000 zum Planen – Steuern – Kontrollieren mittelständischer Fertigungsunternehmen", Kernforschungszentrum Karlsruhe, KfK-PFT 85

Krause, F.-L. 1980 *Systeme der CAD-Technologie für Konstruktion und Arbeitsplanung*, Munich/Vienna

Krause, F.-L. 1983 "Veränderung der Konstruktionstätigheit durch CAD-Systeme", *Produktionstechnisches Kolloquium Berlin, PTK '83*, Munich

Kuby, T. 1980 *Vom Handwerksinstrument zum Maschinensystem, Bildung und Gesellschaft*, vol. 5, TU Berlin

Lang, R. and Hellpach, W. 1922 *Gruppenfabrikation*, Berlin

Lay, G., Boffo, M. and Lemmermeier, L. 1983 "Beurteilung der Wirtschaftlichkeit von CNC-Drehmaschinen unter organisatorischen Gesichtspunkten", Kernforschungszentrum Karlsruhe, KfK-PFT 72

Lay, G. and Lemmermeier, L. 1984 "Werkstattprogrammierung – ja oder nein?" *VDI-Zeitschrift* **126** pp. 595–601

Leibinger, B. 1983 "Die Fertigungstechnik in der Bundesrepublik Deutschland im internationalen Vergleich", *Werkstatttechnik-Zeitschrift für industrielle Fertigung* **73** pp. 277–281

Liese, S. (ed.) 1989 "Werkstattorientierte Programmierverfahren (WOP). Ein Beitrag zur Weiterentwicklung qualifikationsorientierter Produktion", Kernforschungszentrum Karlsruhe, KfK-PFT 138

Lutz, B. and Schultz-Wild, R. (eds.) 1982 *Flexible Fertigungssysteme und Personalwirtschaft*, Frankfurt/New York

Lutz, B. and Schultz-Wild, R. 1983 "Tendenzen und Faktoren des Wandels der Arbeitswelt bei fortschreitender Automatisierung", *Produktionstechnisches Kolloquium Berlin, PTK '83*, Munich

Mallet, S. 1972 *Die neue Arbeiterklasse*, Frankfurt

Malsch, T. 1984 "Erfahrungswissen versus Planungswissen" in U. Jürgens & F. Naschold (eds.) *Arbeitspolitik*, Opladen

Manske, F. and Wobbe-Ohlenburg, W. 1984 "Rechnerunterstützte Systeme der Fertigungssteuerung in der Kleinserienfertigung", Kernforschungszentrum Karlsruhe, KfK-PFT 90

Manske, F. and Wobbe-Ohlenburg, W. 1985 "Fertigungssteuerung im Maschinenbau aus der Perspektive von Unternehmensleitung und Werkstattpersonal", *VDI-Zeitschrift* **127** pp. 395–402 (Teil 1), pp. 457–462 (Teil 2), pp. 489–494 (Teil 3)

Marglin, S.A. 1977 "Was tun die Vorgesetzten? Ursprünge und Funktionen der Hierarchie in der kapitalistischen Produktion". *Technologie und Politik*, vol. 8, Reinbek

Marktübersicht 1984 "CAD/CAM-Marktübersicht", *Markt und Technik* No. 40 pp. 100–124

Martin, T. (ed.) 1984 *Design of Work in Automated Manufacturing Systems*, Oxford/New York/Toronto/Sydney/Paris/Frankfurt

Marx, K. 1969 *Das Kapital*, MEW vols. 23–25, Berlin (DDR)

Mendner, J.H. 1977 *Technologische Entwicklung und Arbeitsprozess*, Frankfurt

Mitrofanow, S.P. 1960 *Wissenschaftliche Grundlagen der Gruppentechnologie*, Berlin (DDR)

Moll, H.H. 1979 "Zeitgerechte Arbeitsgestaltung", *VDI-Zeitschrift* **121** pp. 459–462

Moll, H.H. 1983 "Mehr Produktivität durch weniger Arbeitsteilung", *VDI-Nachrichten* no. 43

Mönig, H. 1980 "Erfahrungen mit einer Fertigungsinsel in der Kunststoffbearbeitung", *Werkstatt und Betrieb* **113** pp. 815–817

Moto-Oka, T. (ed.) 1982 *Fifth Generation Computer Systems*, Amsterdam/New York/Oxford

Müller, W., Patt, P., Schmeink, F. and Zons, K.-H. 1983 "Integrierte Fertigungsunterlagenerstellung", Kernforschungszentrum Karlsruhe, KfK-PFT 60

Mumford, L. 1977 *Mythos der Maschine. Kultur, Technik und Macht*, Frankfurt

Naschold, F. 1984 "Arbeitspolitik – gesamtwirtschaftliche Rahmenbedingungen, betriebliches Bezugsproblem und theoretische Ansätze der Arbeitspolitik" in U. Jürgens & F. Naschold (eds.) *Arbeitspolitik*, Opladen

Neisser, U. 1979 *Kognition und Wirklichkeit. Prinzipien und Implikationen der kognitiven Psychologie*, Stuttgart

Neubauer, G. 1981 "Sozioökonomische Bedingungen der Rationalisierung und der gewerkschaftlichen Rationalisierungsschutzpolitik", dissertation, FU Berlin

Noble, D.F. 1979 *Maschinen gegen Menschen. Die Entwicklung numerisch gesteuerter Werkzeugmaschinen*, Stuttgart

Österreich, R. 1981 *Handlungsregulation und Kontrolle*, Munich

Parnas, D.L. 1985 "Software aspects of strategic defence systems", *Comm. ACM* **28** pp. 1326–1335

Pollard, S. 1973 "Die Fabrikdisziplin in der industriellen Revolution" in R. Braun et al. (eds.) *Gesellschaft in der industriellen Revolution*, Cologne

Popitz, H., Bahrdt, H.R., Jüres, E.A. and Kesting, H. 1957 *Technik und Industriearbeit. Soziologische Untersuchung in der Hüttenindustrie*, Tübingen

Poths, W. 1983 "Den 'CAD/CAM-Zug' nicht verpassen", *Markt und Technik* nos. 16, 17

Puppe, F. 1987 "Diagnostik-Expertensysteme", *Informatik-Spektrum* **10** pp. 283–308

Riehm, U. 1982 "Der Beitrag der Konstruktionswissenschaft zum Einsatz des Rechners in der

Konstruktion, unveröffentlichter Primärbericht", Kernforschungszentrum Karlsruhe

RKW (eds.) 1984 *Integrierte CAD/CAM-Systeme. Entwicklungstrends, Einsatzmöglichkeiten, Auswirkungen*, 2 vols., Rationalisierungskuratorium der deutschen Wirtschaft e.V., Eschborn

Sabel, C.F. 1982 *Work and Politics*, Cambridge/New York/New Rochelle/Melbourne/Sydney

Schaffitzel, W. & Sellmer, U. 1984 "Einführungsstrategien des CAD-Einsatzes. CAD-Anwendung als Aufgabe der Personal- und Organisationsentwicklung", Kernforschungszentrum Karlsruhe, KfK-PFT 89

Scheel, J. 1980 "Einsatz der EDV in der Fertigungsplanung und Fertigungssteuerung", *Arbeitsvorbereitung* **17** pp. 40–42 (Teil 1), pp. 63–66 (Teil 2)

Schlagenhauf, K. and Schaffitzel, W. 1983 "Voraussetzungen und Folgen des CAD/CAM-Einsatzes in der Organisation", *VDI Berichte* no. 492, Düsseldorf

Schulz, H. and Arnold, W. 1983 "Stand und Tendenz beim Einsatz flexibler Fertigungssysteme", *Werkstatt und Betrieb* **116** pp. 61–65

Seliger, G. 1978 "Modularprogramme zur Fertigungssteuerung", *Zeitschrift für wirtschaftliche Fertigung* **73** pp. 199–207

Seliger, G. 1983 *Wirtschaftliche Planung und automatisierte Fertigungssysteme*, Munich/Vienna

Shaiken, H. 1980 "Neue Technologien und Organisation der Arbeit", *Leviathan* pp. 190–211

Sigismund, C. G. 1982 "The structure and strategy of factory automation", SRI International, Business Intelligence Program, research report 661, Menlo Park

Smith, A. 1776 *An Inquiry into the Nature and Causes of the Wealth of Nations*

Sohn-Rethel, A. 1973 *Geistige und körperliche Arbeit*, Frankfurt

Sorge, A. 1985 *Informationstechnik und Arbeit im sozialen Prozess – Arbeitsorgenisation, Qualifikation und Produktivkraftentwicklung*, Frankfurt/New York

Sorge, A., Hartmann, G., Warner, M. and Nicholas, I. 1982 *Mikroelektronik und Arbeit in der Industrie*, Frankfurt

Speith, G., Kittel, T. and Brief, K. 1981 "PPS-Systeme auf dem Prüfstand", *Arbeitsvorbereitung* **18** pp. 113–122

Spinas, P. and Kuhn, R. 1980 "Gruppentechnologie. Ein alternatives Fertigungskonzept", report, Lehrstuhl für Arbeits- und Betriebpsychologie der ETH Zürich

Spinas, P., Troy, N. and Ulich, E. 1983 *Leitfaden zur Einführung und Gestaltung von Arbeit mit Bildschirmsystemen*, Munich

Spur, G. 1979 *Produktionstechnik im Wandel*, Munich

Spur, G. 1983 "Aufschwung, Krisis und Zukunft der Fabrik", *Produktionstechnisches Kolloquium Berlin, PTK '83*, Munich

Spur, G. 1984 "Über intelligente Maschinen und die Zukunft der Fabrik", *Forschung – Mitteilungen der DFG* no. 3, pp. I-VIII

Spur, G. and Ganiyusufoglu, Ö.S. 1983 "Wirtschaftliche Nutzung von flexiblen Fertigungszellen am Beispiel der Drehbearbeitung", *Zeitschrift für wirtschaftliche Fertigung* **78** pp. 176–182

Spur, G. and Mertin, K. 1981 "Flexible Fertigungssysteme, Produktionsanlagen der flexiblen Automatisierung", *Zeitschrift für wirtschaftliche Fertigung* **76** pp. 441–448

Spur, G., Seliger, G. and Eggers, A. 1983 "Kompetenzorientierte Werkstattsteuerung", *Zeitschrift für wirtschaftliche Fertigung* **78** pp. 216–220

Statistisches Jahrbuch der Bundesrepublik Deutschland 1986

Taylor, F.W. 1911 *The Principles of Scientific Management*

Thompson, E.P. 1973 "Zeit, Arbeitsdisziplin und Industriekapitalismus" in R. Braun et al. (eds.) *Gesellschaft in der industriellen Revolution*, Cologne

Thomson, G. 1961 *Die ersten Philosophen*, Berlin (DDR)

Troy, N. and Schüpbach, H. 1984 "Handlungstheoretische Anmerkungen zur Tätigkeit von Produktionsingenieuren", *Zeitschrift für Arbeitswissenschaft* **38** pp. 12–17

Ulich, E., Frei, F. and Baitsch, C. 1980 "Zum Begriff der Persönlichkeitsförderlichen Arbeitsgestaltung", *Zeitschrift für Arbeitswissenschaft* **34** pp. 210–213

Ullrich, O. 1977 *Technik und Herrschaft*, Frankfurt

Ullrich, O. 1979 *Weltniveau. In der Sackgasse des Industriesystems*, Berlin

VDI-Nachrichten 1985 no. 4

VDMA 1982 *Statistisches Handbuch für den Maschinenbau*, Frankfurt

Volpert, W. 1975 "Die Lohnarbeitswissenschaft und die Psychologie der Arbeitstätigkeit" in P. Groskurth & W. Volpert (eds.) *Lohnarbeitspsychologie*, Frankfurt

Volpert, W. 1982 "Der Zusammenhang von Arbeit und Persönlichkeit – Folgerungen für die Arbeitsgestaltung" in J. Albertz (ed.) *Technik und menschliche Existenz*, Freie Akademie, Wiesbaden

Volpert, W. 1984a "Maschinen-Handlungen und Handlungs-Modelle – ein Plädoyer gegen die Normierung des Handelns", *Gestalt Theory* **6** pp. 70–100

Volpert, W. 1984b "Computer und Denken. Machen wir uns selbst zu Maschinen?" *Vortrag auf dem 7. Internationalen Kongress Datenverarbeitung im europäischen Raum*, Vienna

Volpert, W. 1987 "Kontrastive Analyse des Verhältnisses von Mensch und Rechner als Grundlage des System-Designs", *Zeitschrift für Arbeitswissenschaft* **41** pp. 147–152

Warnecke, H.-J., Saak, V. and Häussermann, S. 1979 "Gruppentechnologie und Fertigungszellen", *Werkstattstechnik-Zeitschrift für industrielle Fertigung* **69** pp. 164–166

Warnecke, H.-J., Osman, M. and Weber, G. 1980 "Gruppentechnologie", *Fortschrittliche Betriebsführung/Industrial Engineering* **29** pp. 5–12

Warnecke, H.-J. and Steinhilper, R. 1983a "Neue Entwicklungen zur flexiblen Automatisierung der Teilefertigung", *VDI-Zeitschrift* **125** pp. 853–859

Warnecke, H.-J. and Steinhilper, R. 1983b "Flexible Fertigungssysteme im In- und Ausland", *Technische Zeitschrift für praktische Metallbearbeitung* **77** H. 1, pp. 15–22

Weizenbaum, J. 1977 *Die Macht der Computer und die Ohnmacht der Vernunft*, Frankfurt

Wiendahl, H.-P. 1983 *Betriebsorganisation für Ingenieure*, Munich/Vienna

Williamson, D.T.N. 1967, 1968 "Ein neues Fertigungsverfahren", *Technische Zeitschrift für praktische Metallbearbeitung* **61** pp. 428–439, **62** pp. 39–43

Williamson, D.T.N. 1972 "The anachronistic factory", *Proc. R. Soc. Lond.* **A331** pp. 139–160

Wingert, B., Duus, W., Rader, M. and Riehm, U. 1984 *CAD im Maschinenbau*, Berlin/Heidelberg/New York/Tokyo

Winograd, T. and Flores, F. 1986 *Understanding Computers and Cognition. A New Foundation for Design*, Norwood

Zöller, R. 1980 "Die Fertigung von Bohr- und Frästeilen in einer Fertigungsinsel mit CNC-Maschinen", *Werkstatt und Betrieb* **113** pp. 755–758